NGO Diplomacy

The Influence of Nongovernmental Organizations
in International Environmental Negotiations

NGO Diplomacy

The Influence of Nongovernmental Organizations
in International Environmental Negotiations

edited by Michele M. Betsill and Elisabeth Corell

The MIT Press
Cambridge, Massachusetts
London, England

For information about special quantity discounts, please email special_sales@ mitpress.mit.edu

This book was set in Sabon on 3B2 by Asco Typesetters, Hong Kong.
Printed on recycled paper and bound in the United States of America.

Library of Congress Cataloging-in-Publication Data

NGO diplomacy : the influence of nongovernmental organizations in international environmental negotiations / edited by Michele M. Betsill and Elisabeth Corell.
     p.   cm.
Includes bibliographical references and index.
ISBN 978-0-262-02626-0 (hardcover : alk. paper)—ISBN 978-0-262-52476-6 (pbk. : alk. paper)
1. Environmental policy—International cooperation. 2. Environmental law, International. 3. Negotiation. 4. Non-governmental organizations. I. Betsill, Michele Merrill, 1967– II. Corell, Elisabeth.
GE170.N52   2007
363.7′0526—dc22   2007001898

10   9   8   7   6   5   4   3   2   1

Contents

Foreword

In 1968 Sweden proposed that the United Nations (UN) convene a special conference where the international community could address global environment problems. Perhaps the single most important event in the process of preparation for the upcoming Stockholm Conference was an informal meeting convened in a motel in Founex, Switzerland, in 1971. Attending the meeting were some thirty leading NGOs, environmental experts, and policy leaders. The *Founex Report on Development and Environment* was the outcome of the meeting; its impact on the upcoming Stockholm Conference was unquestionably significant as in many ways it framed the whole outcome of the conference. Maurice Strong, the Secretary General of both the Stockholm and Rio Conferences, said in 1999, "I regard the *'Founex Report on Environment and Development'* as a seminal milestone in the history of the environmental movement."[1] NGOs have been making important contributions ever since.

At the United Nations the term NGO has a particular meaning: "not government," or rather not national governments, as local and regional government associations are also classed as NGOs. The first NGO to be accredited to the UN was in fact the International Chamber of Commerce, perhaps not what NGOs today would see consider as their kind. The Rio Conference in 1992 tried to distinguish among different NGOs by looking at sectors of society that were critical to the implementation of *Agenda 21*. It identified nine "Major Groups," as they were called, or

---

1. *The Hunger Project Millennium Lecture Hunger, Poverty, Population and Environment* by Maurice Strong, April 1999. Available at ⟨http://www.thp.org/reports/strong499.htm⟩.

stakeholders as they are more generally known.[2] This approach has become a framework to enable different stakeholders to become much more involved in implementation than perhaps had been expected in Rio. By involving stakeholders, the whole intergovernmental process became even more complex. At the same time the focus on different stakeholders allowed for much clearer definitions of the roles and responsibilities for monitoring and implementation, beyond governments. NGOs, which have traditionally been thought of as advocacy groups, have found themselves also being subdivided, by themselves or by others, into clusters such as environmental NGOs, community based organizations, and social movements.

This book doesn't attempt to subdivide but to take the term as the UN applies it, and that includes all the above and more. But it is important to understand that just as there has been an increase in the complexity of intergovernmental negotiations due to the increase in the number of governments from 132 in 1972 to 191 by 2002, NGO involvement in this period has also become increasingly complex. As the book notes, in 1972, 250 NGOs were accredited to Stockholm; by the Johannesburg Summit in 2002, there were 3,200 accredited.

The increase in participation of NGOs in global institutions reflects the changing state of our democracy. In 1972, there were only 39 democratic countries in the world; by 2002, there were 139. During those twenty years the changes in Eastern Europe with the fall of the Soviet Union, unthinkable in 1972, and the growing move to democracy in Latin America and Africa were fueled by the growth of civil society within those countries. But what type of democracy? People have become more and more unhappy with the traditional representative democracy—electing individuals with their only involvement being to put a "×" on a piece of paper every four or five years. This questioning at the national level has been accompanied by questioning the democratic deficit in our global institutions, such as the UN and the Bretton Woods institutions. In poll after poll citizens continually said that they trusted the NGOs in their country more than their governments.

2. Indigenous people, trade unions, NGOs, youth and children, women, business and industry, farmers, local authorities, and academics.

In reaction to this, governments have increasingly had to listen to the views of their citizens, often by supporting a particular NGO or NGO position. NGOs can carry considerable political weight. For example, in the United Kingdom the Royal Society for the Protection of Birds has more members than the three main political parties put together. The kind of pressure an NGO can put on governments can indeed persuade them to change policies. This growth of NGO diplomats now plays a significant role in intergovernmental negotiations. Contrasted with government diplomats, NGO diplomats can represent issues that transcend state boundaries, which often affect the global commons. Chip Linder, the organizer of the Global Forum at Rio, said: "it became the first international experiment in democratizing intergovernmental decision making." One question that will become more and more important in the next decade is whom do NGO diplomats represent? Are their governance structures transparent and accountable? All questions for another book.

NGO Diplomacy is therefore a timely and an extremely important contribution to understanding what impact NGOs have had in intergovernmental negotiations on environment and sustainable development, and how they achieved that impact. Are there lessons here for future generations on what approach can best ensure governments adopt their positions?

Challenges

This volume presents an analytical framework for the study of NGO diplomacy that takes into account the effects of NGOs on both decision-making processes and negotiation outcomes. The framework provides a basis for conducting systematic comparative analyses. Contributors use the framework to examine the role of NGO diplomats in negotiations on climate change, biosafety, desertification, whaling, and forests.

The framework is fascinating for someone who spends a lot of their time trying to influence negotiators. It recognizes that there are key stages to influence the negotiating process such as issues framing, agenda setting, and understanding the position of key actors. Not all the work is done at the global level; many NGOs work at persuading their governments in capitals before they attend international meetings. Ultimately the book recognizes that the measures of success for NGOs should be:

Can the outcomes be tracked back to positions that NGOs took, and does the text reflect those positions?

Of course, often NGO work is undertaken away from the meeting rooms, and not always in published form. Those NGOs working inside the system can find it counterproductive to publish the position they want governments to take. Stakeholder Forum (the organization of which I am Executive Director) often uses dinners or other means to create a government support group on a particular issue. We also send suggestions directly to governments before the meeting and discuss, over email or telephone or in person, why a particular position should be taken. The use of interviews with NGOs, and not just relying on published material, is very important to understanding what has happened in a particular negotiation.

On a more depressing note, not all is rosy with regards to the involvement of NGOs at the UN since 9/11. There have been questions raised about the possibility of terrorists hiding in NGO clothes, or NGOs knowingly, or unknowingly, financing terrorists.

As we have seen, with the growth in the involvement of NGOs in the UN multilateralism has also diminished, seemingly attacked from many fronts. NGOs have been one of the strongest supporters of the UN in the last decade as they see the need for international legal framework for preserving our global commons.

Over the last five years there has also been a reduction of funding for NGOs. Often donor governments funded NGOs in nondemocratic countries as a way of ensuring that funding reached the poorest people in society. As democracy has grown, governments have shifted funding to national budgets and away from projects. This is support for nation building.

I strongly recommend this book to anyone who is or who wants to be involved or anyone who just wants to understand how international negotiations are conducted, and the roles NGOs have in such negotiations. As the editors say, "there is more work to be done." I look forward to seeing how NGOs use the framework as part of their own development.

Felix Dodds
San Sebastian

Acknowledgments

Our collaboration began in 1999, and we have received a great deal of help and support along the way. We are extremely grateful to Jenny Wahren for her research assistance. We also wish to thank the Swedish Research Council and the National Science Foundation (SES-031865) for funding a 2003 workshop, which was graciously hosted by the Swedish Institute of International Affairs. Gunilla Reischl and Elisa Peter provided invaluable assistance in coordinating the workshop, and Tore Brænd, Felix Dodds, Elisa Peter, Beatriz Torres, Stacy VanDeveer, and Paul Wapner offered excellent feedback on the draft chapters. Colorado State University and the Fulbright Association made it possible for Elisabeth to spend three months in Colorado in 2004. Over the years we have received considerable support from members of the International Studies Association's Environmental Studies Section, who provided us a forum for discussing the project as it evolved and gave us valuable feedback and encouragement. We also thank Clay Morgan at The MIT Press for his enduring patience and support. Finally, we are grateful to our families for helping us keep it all in perspective.

Contributors

Steinar Andresen is a senior research fellow at the Fridtjof Nansen Institute (FNI) where he has spent most of his time. He was a visiting scholar at the University of Washington in 1987–1988, a part-time associate with the International Institute of Applied Systems Analysis (IIASA) in 1994–1996, a visiting scholar at Princeton University in 1997–1998, and a professor of political science at the Department of Political Science, University of Oslo in 2002–2006. He has worked on law of the sea issues, international environmental, and resources regimes, including the whaling regime. He has published five books, been guest editor of three special issues of international journals (both mostly with co-authors), and has published extensively in international journals.

Michele Betsill is Associate Professor of Political Science at Colorado State University where she teaches courses in international relations, global environmental politics, and research methods. She was an Affiliate Scientist with the Institute for the Study of Society and the Environment at the National Center for Atmospheric Research in 2004–2007. Her research focuses on the governance of global environmental problems, especially related to climate change. She is author and co-author of numerous book chapters and articles. She is co-author (with Harriet Bulkeley) of *Cities and Climate Change: Urban Sustainability and Global Environmental Governance* (Routledge, 2003) and co-editor (with Kathryn Hochstetler and Dimitris Stevis) of *Palgrave Advances in International Environmental Politics* (Palgrave, 2006).

Stanley Burgiel is the Senior International Policy Analyst for the Nature Conservancy's Global Invasive Species Initiative where he focuses on the Convention on Biological Diversity, free trade agreements, and the development of national systems to prevent new introductions of invasive species. He has also worked on multilateral policy issues for Defenders of Wildlife and the Biodiversity Action Network. Stas focused on biosafety issues while a writer and editor for the *Earth Negotiations Bulletin*, where he also covered international negotiations related to biodiversity and plant genetic resources. He has consulted for a range of environmental organizations, including the International Institute for Sustainable Development, the Global Forest Policy Project, and the World Foundation for Environment and Development. Stanley has authored, co-authored, or edited more than fifty articles, monographs, and meeting reports on international policy issues.

Elisabeth Corell's research focuses on international decision making for sustainable development, with particular interest in the role of experts and scientific advisors as well as actors who represent practical or experience-based knowledge. Elisabeth has covered international negotiations on desertification and meetings of the United Nations Environment Program's governing council for the *Earth Negotiations Bulletin*. She is co-editor (with Angela Churie Kallhauge and Gunnar Sjöstedt) of *Global Challenges: Furthering the Multilateral Process for Sustainable Development* (Greenleaf, 2005).

David Humphreys is Senior Lecturer in Environmental Policy at the Open University. He is the author of *Forest Politics: The Evolution of International Cooperation* (Earthscan, 1996) and *Logjam: Deforestation and the Crisis of Global Governance* (Earthscan, 2006). He is co-editor (with Alan Thomas and Susan Carr) of *Environmental Policies and NGO Influence: Land degradation and sustainable resource management in sub-Saharan Africa* (Routledge, 2001). He has attended five international negotiating and expert group meetings on forests as either an NGO observer or member of a UK government delegation. He was an adviser to the World Commission on Forests and Sustainable Development and is currently serving on the Scientific Advisory Board of the European Forest Institute.

Tora Skodvin is a senior research fellow at Center for International Climate and Environmental Research—Oslo (CICERO). Her research interests include international climate negotiations, with a particular focus on the role of non-state actors in general and scientific communities in particular. Book publications include *Structure and Agent in the Scientific Diplomacy of Climate Change* (Kluwer, 2000) and *The Oil Industry and Climate Change* (with Jon Birger Skjærseth, Manchester University Press, 2003).

Acronyms

AOSIS	Alliance of Small Island States
ASSINSEL	International Association of Plant Breeders
BIO	Biotechnology Industry Organization
BSWG	Biosafety Working Group
CAN	Climate Action Network
CBD	Convention on Biological Diversity
CCD	Convention to Combat Desertification
CEE	Central and Eastern Europe
CITES	Convention on International Trade in Endangered Species
COP	Conference of the Parties
CSD	Commission on Sustainable Development
EDF	Environmental Defense Fund
ENGI	Environmental nongovernmental individual
ENGO	Environmental nongovernmental organization
ESA	Endangered Species Act
ExCOP	Extraordinary Conference of the Parties
FERN	Forests and the European Union Resource Network
FOE	Friends of the Earth
FPS	Fauna Protection Society
G-77	Group of 77 Developing Countries
GATS	General Agreement on Trade in Services
GATT	General Agreement on Tariffs and Trade
GCC	Global Climate Coalition

GFPP	Global Forest Policy Project
GHG	Greenhouse gas
GIBiP	Green Industry Biotechnology Platform
GIC	Global Industry Coalition
GM	Genetically modified
GMO	Genetically modified organism
ICRW	International Convention for the Regulation of Whaling
IFF	Intergovernmental Forum on Forests (1997–2000)
IGO	Intergovernmental organization
IIED	International Institute for Environment and Development
IISD	International Institute for Sustainable Development
ILO	International Labor Organization
IMF	International Monetary Fund
INCD	Intergovernmental Negotiating Committee on Desertification
IPED	International Panel of Experts on Desertification
IPF	Intergovernmental Panel on Forests
ISA	International Studies Association
ITTA	International Tropical Timber Agreement
ITTO	International Tropical Timber Organization
IUCN	International Union for the Conservation of Nature
IWC	International Whaling Commission
LMO	Living modified organism
LMO-FFP	Living modified organism intended for direct use as food, feed or processing
NAMMCO	The North Atlantic Marine Mammal Commission
NGO	Nongovernmental organization
OPEC	Organization for Petroleum Exporting Countries
PACD	Plan of Action to Combat Desertification
RIOD	Le Réseau d'ONG sur la Désertification et la Sécheresse
RMP	Revised Management Procedure
SPS	Sanitary and Phytosanitary Standards

TFRK	Traditional forest-related knowledge
TWN	Third World Network
UNCED	United Nations Conference on Environment and Development
UNDP	United Nations Development Programme
UNEP	United Nations Environment Programme
UNFCCC	United Nations Framework Convention on Climate Change
UNFF	United Nations Forum on Forests
WTO	World Trade Organization
WWF	World Wide Fund for Nature/World Wildlife Fund (in North America)

1

Introduction to NGO Diplomacy

Michele M. Betsill and Elisabeth Corell

The modern era of international decision making on the environment and sustainable development formally began with the 1972 United Nations Conference on the Human Environment, held in Stockholm. Representatives of more than 250 nongovernmental organizations (NGOs) attended the Stockholm Conference, representing constituencies bound by common values, knowledge, and/or interests. These NGOs served as technical experts, helped develop the rules for NGO participation, participated in plenary sessions and committee meetings, and engaged in several parallel forums designed to strengthen their connections with one another. Willetts (1996b: 57) views Stockholm as a watershed event in terms of NGO involvement in global governance, marking the beginning of a "slow yet steady liberalization of the NGO system occurring over the following two decades."

Since Stockholm, NGO involvement in international decision-making processes related to the environment and sustainable development has escalated, as demonstrated by their participation in the two subsequent global conferences. More than 1,400 NGOs were accredited to the 1992 United Nations Conference on Environment and Development, held in Rio de Janeiro, and more than 25,000 individuals from 167 countries participated in the parallel Global Forum, where NGOs negotiated alternative treaties and engaged in extensive networking (Chatterjee and Finger 1994; Dodds 2001; Kakabadse and Burns 1994; Morphet 1996; Willetts 1996b). One of the greatest achievements of the Rio Conference was *Agenda 21*, the action plan for sustainable development in the twenty-first century, which recognized NGOs as partners in the global struggle to promote sustainable development. In 2002, more than 3,200 organizations were accredited to the World Summit on Sustainable Development in Johannesburg, where NGOs were central to the creation of partnerships for sustainable development (Gutman 2003; Speth 2003).

The dramatic increase in the number of NGOs over the past century has been well documented, as has the fact that these organizations increasingly participate in international political processes. Academic interest in the role of these actors in global environmental politics has exploded since the early 1990s, and a growing body of evidence indicates that NGOs influence government decisions to develop domestic policies to protect natural resources and to negotiate international treaties, as well as how individuals perceive environmental problems (see Betsill 2006). Despite mounting evidence that NGOs make a difference in global environmental politics, the question of under what conditions NGOs matter generally remains unanswered.

This volume addresses this question in the realm of international environmental negotiations. We contend that the increased participation of NGOs in these political processes reflects broader changes in the nature of diplomacy in world politics. In international relations scholarship, diplomacy is often viewed as something that states do; an important aspect of statecraft and foreign policy (e.g., Magalhães 1988). Alternatively, Sharp (1999) argues that diplomacy is better understood in terms of representation; diplomats are actors who act on the behalf of a clearly identified constituency. We find that Sharp's definition better captures the reality of multilateral negotiations on the environment and sustainable development. As the contributions in this volume demonstrate, international environmental negotiations cannot be understood in terms of inter-state diplomacy. Rather, these processes involve myriad actors representing a diversity of interests. In multilateral negotiations on the environment and sustainable development, NGO representatives act as diplomats who, in contrast to government diplomats, represent constituencies that are not bound by territory but by common values, knowledge, and/or interests related to a specific issue (see Starkey, Boyer, and Wilkenfeld 2005).

To the extent that NGO diplomacy has been considered in the past, the emphasis has often been on unofficial acts, such as hosting foreign visitors or participating in cultural exchanges or scientific meetings (sometimes referred to as "citizen" or "track-two" diplomacy) (see National Council for International Visitors 2006; Starkey, Boyer, and Willkenfeld 2005). However, these discussions typically treat NGO diplomacy as something that occurs outside the realm of formal, inter-state politics. In contrast, the contributions in this volume illuminate the ways that NGOs engage directly in one of the most traditional diplomatic activities—formal international negotiations. In each of our cases, NGO diplomats perform many of the same functions as state delegates: they represent the interests of their constituencies, they engage in information exchange, they negotiate, and they provide policy advice (Aviel 2005; Jönsson 2002).

This volume presents an analytical framework for the study of NGO diplomacy that takes into account the effects of nongovernmental organizations on both negotiation processes and outcomes and provides a basis for conducting systematic comparative analyses. Most current research consists of individual case studies, where scholars rely on different measures of NGO influence, different types of data, and different methodologies. As a result it is difficult to make assessments about where NGOs have had more or less influence and to examine the factors that may lead to variation in NGO influence across cases. In this volume, contributors use the framework to examine the role of NGO diplomats in negotiations on climate change, biosafety, desertification, whaling, and forests. Within these cases many different types of NGOs are considered—environmental, social, scientific, and business/industry organizations. These analyses demonstrate that it is possible to make qualitative judgments about levels of NGO influence and that comparison across the cases allows scholars to identify factors that explain variation in NGO influence in different negotiating situations.

In this introductory chapter we define what we mean by NGOs and clarify our focus on international negotiations. We then discuss the need for a systematic approach to the study of NGO influence in international environmental negotiations and outline the strategy we have used to conduct such research in this project. We conclude with an overview of the remaining chapters in the volume.

What Are NGOs?

Scholars and practitioners use the term NGO to refer to a wide range of organizations, which are often differentiated in terms of geographic scope, substantive issue area, and/or type of activity. Some authors specifically examine international NGOs working in at least three countries, while others focus on national or local grassroots organizations. Still others emphasize the various networks formed by these organizations. Studies of international environmental negotiations routinely highlight the involvement of environmental NGOs (ENGOs) as well as scientific organizations and NGOs representing business and industry interests. Finally, some scholars differentiate between NGOs based on the character of their primary activities: advocacy, research, and outreach.

In this project, the term "NGO" refers to a broad spectrum of actors from advocacy groups rooted in civil society to privately held multinational corporations and trade associations to research-oriented bodies that participate in international environmental negotiation processes using the tools of diplomacy.[1] We draw on Oberthür et al.'s (2002) thorough review of the legal and academic literature on NGOs, which identified three minimum criteria that are used in the accreditation process to determine who may participate in international policy making processes and thus to define an NGO. According to this study, an NGO is an organization that (1) is not formed by intergovernmental agreement, (2) has expertise or interests relevant to the international institution, and (3) expresses views that are independent of any national government. This is consistent with how the term is used in the UN, which also excludes organizations that advocate violence, are political parties, and/or do not support UN objectives (Oberthür et al. 2002; Willetts 1996b).

For the purposes of the present study, this broad usage of the term NGO is appropriate for at least two reasons. First, as stated above, it reflects the usage within the UN system, which covers the majority of international institutions in which multilateral negotiations related to the environment and sustainable development take place. Second, all NGO representatives can be distinguished from state diplomats in that they do not represent territorially defined interests. We recognize the diversity of actors that fall within this definition and have encouraged the contributors to make distinctions between types of NGOs (e.g., environmental groups vs. industry associations) as they see fit. However, we did not wish to exclude a priori any type of NGO, since the purpose of this project was to explore the significance of NGO diplomacy, broadly defined, on international environmental negotiations. We recognize, however, that there may be important differences between types of NGOs that affect whether and how they exert influence. The framework we develop to analyze NGO influence in international environmental negotiations may help illuminate these differences. We address the importance of the distinctions between NGOs in the conclusions and suggest areas for future research on this important question.

Why International Negotiations?

International negotiations are one political *arena* in which NGOs attempt to shape policy making related to the environment and sustainable development (see Betsill 2006). Other arenas include (this is not an exhaustive list): domestic policy making, the formation of global civil society, and decision making of private actors (e.g., corporations). While NGO activities in all of these political arenas may have implications for the global governance of the environment and sustainable development, we argue that each of these arenas is likely to involve different political dynamics that in turn shape the ways that NGOs participate, the goals they pursue, the strategies they use and the likelihood that they will achieve those goals (Betsill and Corell 2001).

Unfortunately, much of the current literature tends to treat all studies related to NGOs in the area of environment and sustainable development as a single body of research, without differentiating between these different arenas of activity. While NGOs may be central in the development of a global civil society, it is entirely possible that they are less successful in shaping new international institutions to address environmental issues. Scholars need to employ a multifaceted view of the role of NGOs and the arenas in which they participate in world politics. At the same time there is great demand for general conclusions about

NGO influence in international politics. It would be also useful to be able to consider whether NGOs are generally more influential in particular arenas, and if so, why.

The purpose of this project is to better understand these dynamics within one arena—international environmental negotiations. We examine negotiations aimed at creating a new agreement outlining general principles, commitments, and/or decision-making procedures as well as post-agreement negotiations that address questions of implementation and/or new conflicts that arise under an existing treaty (Spector and Zartman 2003). International negotiations are a particularly interesting arena in which to consider NGO influence since they are largely the domain of states. As UN members, only states have formal decision-making power during international negotiations. They establish rules for who may participate and the nature of that participation (e.g., through formal interventions or by directly engaging in floor debate), and ultimately it is states that vote on whether to adopt a particular decision. In contrast, NGOs often participate in these processes as observers and have no formal voting authority, making it difficult for NGO diplomats to influence the negotiating process. Thus findings of NGO influence in international environmental negotiations present an interesting empirical puzzle.

In this volume we specifically analyze NGOs who attend international negotiations for the purpose of influencing those negotiations. Many NGOs attend negotiations to take advantage of the opportunities to network with other NGOs; they show very little interest in engaging in NGO diplomacy (Friedman, Hochstetler, and Clark 2005). While the development of such networks may have significant implications for global environmental politics more broadly, we are primarily interested in the more immediate effects of NGO diplomacy on specific negotiating situations.

We wish to clarify two points related to our understanding of multilateral negotiations. First, negotiation processes and outcomes are shaped by more than just what happens during isolated, two-week formal negotiating sessions.[2] NGO diplomats may influence multilateral negotiations during the pre-negotiation/agenda-setting phase, so it is important to consider how the negotiations came about in the first place. In addition NGOs may influence the negotiation process during formal intersessional meetings, through domestic channels and/or in more informal settings as well. Therefore, in assessing the influence of NGO diplomats in international negotiations, we have encouraged contributors to consider all activities related to multilateral negotiations, not just those that occur during the official two-week sessions.

Second, our conception of political arenas should not be confused with levels of analysis. The dynamics within the political *arena* of international negotiations are shaped by things that happen at different *levels*, including the domestic level.[3] To the extent that NGOs engage in activities within a domestic context that are clearly *targeted* at influencing international negotiations, these activities should be considered in the analysis of NGO diplomacy.

A Systematic Approach

Despite mounting evidence that NGOs make a difference in global environmental politics, the question of under what conditions they matter remains unanswered. Specifically, it is difficult to draw general lessons about the role of NGO diplomacy in international negotiations on the environment and sustainable development because the current literature suffers from three weaknesses.[4] First, as noted above, there is a tendency to treat all studies related to NGOs in the environmental issue area as a single body of research without distinguishing between the different political arenas in which they operate. It is important not to collapse conclusions in the literature about these different spheres of activity. Students of NGOs need to employ a multifaceted view of the role of NGOs.

Second, there is a surprising lack of specification about what is meant by "influence" and how to identify NGO influence in any given political arena (two notable and commendable exceptions are Arts 1998 and Newell 2000). Progress in our understanding of the conditions of NGO influence in international environmental negotiations depends on more careful consideration of what we mean by NGO influence and how influence might be identified. While we recognize that defining influence can be a complicated matter, it is highly important because it forces analysts to think carefully about the types of evidence needed to indicate NGO influence. Without a clear understanding of what is meant by influence, scholars often appear to be presenting evidence on an ad hoc basis. As a result such studies run the risk of overdetermination as scholars look for any possible sign that NGOs made a difference in a given political process while ignoring evidence to the contrary. In other words, defining influence has implications for the robustness of research findings. Moreover lack of consistency in the types of evidence used to indicate NGO influence in international environmental negotiations makes it difficult to compare the role of NGO diplomats across cases, to make assessments about where NGOs have had more or less influence, and to examine the factors that may lead to variation in NGO influence across cases.

Another problem associated with the failure to define influence is that the evidence presented may not be an appropriate proxy for NGO influence. If NGO diplomats truly influence international environmental negotiations, then it should be possible to observe the effects of that influence (King, Keohane, and Verba 1994). Scholars frequently rely on evidence regarding NGO *activities* (e.g., lobbying, submitting information or draft decisions to negotiators on a particular position), their *access* to negotiations (e.g., number of NGOs attending negotiations and the rules of participation) and/or NGO *resources* (e.g., knowledge, financial and other assets, number of supporters and their particular role in negotiations). However, these types of evidence primarily tell us *how* NGOs engage in international environmental negotiations but do not give us information on the subsequent *effects*.

Third, most studies stop short of elaborating the causal linkages between NGO activities and outcomes. Gathering evidence of NGO influence in a more systematic fashion is clearly an important first step to enhancing our understanding of how and under what conditions NGO diplomats matter in international environmental negotiations. However, researchers still run the risk of confusing correlation with causation. If a particular proposal for discussion or wording in the agreement text corresponds to views of NGOs, does that necessarily reflect the success of NGO diplomacy? It could be the case that other actors involved in the negotiations were promoting similar views. Plausibility claims can be strengthened by linking NGO participation and influence in international environmental negotiations.

In sum, progress in understanding under what conditions NGOs matter can be achieved by more carefully recognizing the distinct political arenas in which NGOs operate, by defining what we mean by NGO "influence," and by elaborating the processes by which NGO diplomats influence multilateral environmental negotiations. In this volume we further theoretical development on the role of NGOs in global environmental politics by proposing an analytical framework for assessing their influence in one sphere of activity—international environmental negotiations. The framework, which takes into account the effects of NGO diplomats on both negotiation processes and outcomes, provides a basis for conducting systematic, comparative analyses, which in turn allow us to make some claims about the conditions under which NGOs matter.

Research Design

This volume is the culmination of a project begun in 1999. The objectives of the project are twofold: (1) to develop methodologies for strengthening findings of NGO influence in international environmental negotiations, and (2) through comparative analysis, to identify a set of conditioning factors that shape the ability of NGO diplomats to influence such negotiations. At the core of the project is an analytical framework for assessing NGO influence in international environmental negotiations, which was originally published in 2001 (Betsill and Corell 2001; Corell and Betsill 2001). Shortly thereafter, project participants began developing case studies to both test and refine the framework as a tool for assessing NGO influence and to begin discussions of the conditioning factors that shape NGO influence.

The cases have been selected based on the availability and interest of scholars with significant prior knowledge of NGO diplomacy in international environmental negotiations. Three cases (climate change, biosafety, and desertification) examine single agreement negotiations over a fairly short period of time. The other two cases (whaling and forests) analyze several negotiations on a single issue over a decade or more and often in different institutional contexts. These latter cases provide the opportunity to consider how NGO influence changes over time, across institutional fora, and/or as negotiations pass through different phases. The cases cover negotiations of initial agreements as well as post-agreement negotiations focused on how to achieve an agreement's goals and address ongoing or new conflicts that arise (Spector and Zartman 2003). The cases are heavily weighted toward natural resource issues as opposed to pollution.

Regarding our first objective—developing methodologies for analyzing NGO influence in international environmental negotiations—our approach to case selection is unproblematic. The cases are appropriate in that in each instance NGOs were actively engaged in international negotiations, giving us the opportunity to evaluate the utility of the proposed analytical framework for assessing NGO influence in this particular political arena. We are, however, more limited in terms of our second objective—to identify a set of conditioning factors that shape the ability of NGO diplomats to influence such negotiations. Our opportunistic approach to case selection precluded us from engaging in a "theory-testing" exercise in our cross-case analysis, since we made no determination about the appropriateness of the cases at the outset (see George and Bennett 2005). Instead, we took a more heuristic approach whereby each of the case authors inductively identified the key conditioning factors that enabled or constrained the ability of NGO diplomats to influence international negotiations in their respective issue areas. We then conducted a plausibility probe by examining eight of these factors across the cases to identify those factors warranting further research. This should not been seen as an exhaustive list of the factors that might shape the ability of NGOs to influence international environmental negotiations; the general literature on NGOs suggests many others that need to be analyzed more systematically (see chapter 2).

We urge readers to exercise caution in generalizing our findings beyond these case studies. The majority of our cases examine environmental NGOs; thus we are limited in what we can say about differences in the conditions under which different types of NGOs are likely to influence international environmental negotiations. In addition more than half of the cases used in the cross-case analysis are related to forestry negotiations. We strongly encourage scholars to subject the issues raised in this volume as well as hypotheses from the broader literature to rigorous analysis based on a more careful selection of cases.

The framework and case studies have been presented at two annual meetings of the International Studies Association where we received many helpful comments from fellow academics. In August 2003 we held a workshop in Stockholm, Sweden, which brought together project participants and NGO practitioners with extensive experience in the negotiation processes under analysis.[5] The Stockholm Workshop provided an excellent opportunity to ground the scholarly research on NGO influence in international negotiations in the actual experience of NGO diplomats. The practitioners offered many valuable insights that might not otherwise be available to academic researchers. Prompted by the framework, practitioners also had the rare opportunity to reflect on their own efforts and their organizations' impact on international environmental negotiations. Through the dialogue that took place over the weekend, members of both communities gained a better understanding of one another.

Overview

Chapter 2 elaborates the analytical framework at the core of the project. The framework provides a basis for conducting systematic comparative analysis by addressing many of the weaknesses in the current literature noted above. It begins by identifying two dimensions of NGO influence: participation in international negotiations and the subsequent effects on the behavior of other actors (e.g., states). Scholars are then encouraged to gather data on these two dimensions from a variety of sources, including primary and secondary documents, interviews, and where possible, participant observation. Using the analytical techniques of process tracing and counterfactual analysis, researchers should identify whether and how NGO diplomats shaped both the negotiation *process* (through issue framing, agenda setting, and/or by shaping the positions of key states) as well as the final *outcome* (procedural and substantive elements of the final text). By considering the range of effects NGO diplomats may have on international environmental negotiations, scholars can make a qualitative assessment of the overall influence of NGOs. Results may range from low levels of influence, where NGO diplomats participate but have little effect on either the negotiation process or outcome, to high levels of influence, where NGO diplomacy is linked to effects on both process and outcome. Chapter 2 concludes with a discussion of conditioning factors that make NGO influence more or less likely in any given negotiating context.

The empirical chapters apply the framework in five case studies of international environmental negotiations. Although the authors exhibit different styles in using the framework, each chapter consists of a detailed narrative in which the authors present evidence related to NGO participation and subsequent effects, assess their overall influence on negotiation processes and outcomes, and identify several factors seen to have either enabled or constrained NGO diplomats in their efforts to influence the negotiations.

In chapter 3, Michele Betsill analyzes the role of environmental NGOs in the first phase (1995–1997) of the Kyoto Protocol negotiations on global climate change. Betsill examines negotiations on the issues of targets and timetables, emissions trading, and sinks and assesses whether NGOs were successful in achieving their goals on each of these issues. Overall, she concludes that the environmental community had a moderate level of influence on the negotiations. They had little effect on the outcome of the negotiations; NGO positions on each of these issues are not reflected in the Kyoto Protocol text. Environmental NGOs did, however, shape the negotiation process by working behind the scenes to raise concerns about issues on the negotiation agenda and to influence the positions of key states. Betsill identifies NGO coordination and creativity as important enabling factors related to NGO influence. At the same time significant contention over the economic aspects of controlling greenhouse gas emissions, a focus on finding technological solutions, and the expectation that the Protocol would include binding commitments limited the political space available to the environmental community to achieve their objectives.

Stanley Burgiel compares the influence of environmental and industry NGOs in the negotiations of the Cartagena Protocol on Biosafety (1995–2000) in chapter 4. Burgiel focuses his analysis on four major issues in the negotiations: the agreement's scope, trade-related concerns, decision-making criteria, and exporter responsibilities. He concludes that both groups had moderate influence on the negotiations, with greater success in shaping the negotiation process than outcome. Environmental and industry NGOs both exerted influence by shaping the position of (different) key states and shaping the agenda. Interestingly Burgiel finds that environmental NGOs often focused on getting or keeping issues on the agenda, while industry NGOs worked diligently to keep issues off the agenda. He argues that alliances with key states were a crucial factor enabling non-state actors to exert influence in the Cartagena Protocol negotiations.

In chapter 5, Elisabeth Corell examines the influence of environmental and social NGOs in the negotiation of the United Nations Convention to Combat Desertification (1993–1997). These groups worked together to encourage the use of a "bottom-up approach" to implementation, and to ensure that the Convention recognized the social and economic consequences of land degradation for affected populations and provided additional resources for dryland management projects. She contends that NGOs had a high level of influence on the Convention negotiations as their activities had observable effects on both the negotiation process and outcome. Corell finds several instances where NGO proposals made their way into the treaty text and notes that NGOs were effective in securing participation rights in the negotiations, which in turn gave them the opportunity to shape the negotiation agenda. She attributes the high level of NGO influence in this case to three factors: the link between the bottom-up approach and NGO participation in implementation, the homogeneity of NGOs participating in the negotiations, and the fact that NGO participation was actively encouraged by the negotiators.

In chapter 6, Steinar Andresen and Tora Skodvin assess the influence of the scientific community and environmental NGOs in two major negotiations within the International Whaling Commission: the adoption of a new management procedure in 1974 and a ban on commercial whaling in 1982. Andresen and Skodvin assess non-state actor influence through two channels: directly at the international level and indirectly via the domestic channel. They contend that the scientific community had a moderate degree of influence on the 1974 decision to adopt a new management procedure, primarily by framing the debate at the international level through the provision of technical information. This influence was facilitated by the political demand for advanced knowledge on whale stocks and scientific consensus on the need for more restrictive management procedures. Moreover there were no other non-state actors competing for influence as the environmental community had not yet mobilized on the whaling issue. By the 1980s, the environmental community had become an active player in whaling negotiations, and Andresen and Skodvin argue they had a high level of influence on the 1982 moratorium decision. Factors that helped environmental NGOs achieve their goal include heightened public concern, which opened up important domestic channels of NGO influence in key states, and the availability of significant financial resources. Andresen and Skodvin contend that the increased influence of the environmental community came as the influence of the scientific community declined, largely due to polarization among scientists on the need for a moratorium.

Finally, David Humphreys traces attempts by environmental NGOs to influence international negotiations on forests from the mid-1980s through 2001 in several different institutional contexts in chapter 7. He examines forest negotiations at the United Nations Conference on Environment and Development, under the auspices of the Commission on Sustainable Development, and the consultation process that led to the creation of the United Nations Forum on Forests. He also considers two negotiation processes on forest products, namely negotiations on the international trade of tropical timber in the International Tropical Timber Organization and the discussions on forest products that took place within the World Trade Organization in the late-1990s. Overall, Humphreys concludes that NGOs had a high level of influence on international forest negotiations during this period, although their influence on negotiation processes in the different institutional contexts varied. He argues that the prospects for influence increased when NGOs shaped the negotiation agenda early on. At the same time, the deep North–South division on forest issues has often limited the political space available to NGOs during the negotiations. Humphreys concludes by arguing that environmental NGOs' most important contribution occurred over time rather than in any specific negotiation; they have succeeded in reframing the issue of forest conservation from a purely economic issue to an ecological and human rights one.

Chapter 8 returns to the project's two main objectives. We begin by reflecting on the analytical framework's utility in strengthening claims of NGO influence in international environmental negotiations. The empirical chapters demonstrate that the framework can be used to strengthen claims of NGO influence by elaborating some of the causal links between NGO activities and observed effects on negotiating processes and outcomes. We conclude that the framework works best for analyzing NGO influence in discrete sets of negotiations rather than in multiple negotiations in an issue area as assessments of NGO influence in such cases may be overdetermined by aggregating data over a longer period of time. The cases also demonstrate that it is possible to make qualitative judgments about levels of NGO influence, differentiating among low, moderate, and high levels of influence. However, we found that it was not always straightforward which category was most appropriate in any given case. International environmental negotiations cover numerous highly technical issues simultaneously, and NGOs may influence negotiation processes and/or outcomes on some issues but not others. In the future we suggest that analysts may find it more useful to assess NGO influence at the level of individual issues rather than on the overall negotiations.

Next, we discuss how comparison across cases allows identification of factors that explain variation in NGO influence in different negotiating situations. As mentioned above, we asked the case authors to identify the key factors that enhanced or constrained the ability of NGO diplomats to influence international environmental negotiations. We conducted a cross-case analysis of the eight factors that came up most frequently, resulting in a number of findings warranting future research:

• *NGO coordination* has a neutral effect on influence. In our cases, NGO diplomats achieved all levels of influence under conditions of coordination, and in one of our cases of relatively high influence, NGOs had no coordinated position or strategy.

• NGO influence does not necessarily decline as *rules of access* become more restrictive because NGO diplomats are often quite innovative in finding alternative strategies. At the same time opportunities for NGO influence may be enhanced when state delegates and convention secretariats take steps to actively facilitate NGO participation in the negotiations.

• For environmental NGOs, influence in the early *stages of negotiations* (e.g., debates over the negotiation agenda) may be necessary though by no means sufficient for achieving influence in later stages (e.g., debates over the specifics of the agreement text). This finding may not hold, however, for industry NGOs.

• Higher levels of NGO influence are more likely when the *political stakes* of the negotiations are relatively low, as in negotiations over non-binding principles and/or framework agreements with few demands for behavioral change. NGO diplomats can enhance their ability to influence negotiations with higher political stakes by developing close personal relationships with state diplomats and/or convincing negotiators that NGOs are essential partners in achieving the agreement's objectives.

• *Institutional overlap* offers opportunities for NGO diplomats to influence a given negotiation process indirectly by exerting influence in a related institutional setting. However, overlap with the World Trade Organization and the international trade regime may constrain the ability of environmental NGOs to exert influence while enhancing opportunities for NGOs representing business/industry interests.

• *Competition from other NGOs* is not necessarily a constraining factor because NGO influence in international environmental negotiations is not a zero-sum game. Different types of NGOs often focus on different issues within the negotiations so that each may exert influence on particular issues without taking away from the others.

• Opportunities for influence appear to be enhanced when NGOs form *alliances with key states*. However, such alliances may be less useful in negotiations where states are highly polarized (e.g., along North–South lines).

• Where there is a high *level of contention* over entrenched economic interests, environmental NGOs may have greater difficulty exerting influence on the negotiations. In contrast, contention over the economic aspects of an environmental problem may open up opportunities for business/industry NGOs to influence the negotiations.

The volume concludes by discussing the broader contributions of the project. The cases demonstrate the changing nature of diplomacy in the international system, highlighting the ways that NGO diplomats participate in and influence international negotiations on the environment and sustainable development. We also consider the relationship between the findings of this project and current debates about restructuring the existing system of global environmental governance, specifically the role of NGOs in the realm of international decision making, and efforts to democratize global governance.

Notes

1. This differentiates our study from the social movements literature, which analyzes networks and organizations that tend to mobilize their constituents through protest or disruptive action and are interested in opening up opportunities for mass participation (Khagram, Riker, and Sikkink 2002; Yearley 1994).

2. Thanks to the participants at the August 2003 Stockholm Workshop for pushing us on this point.

3. We are particularly grateful to Tora Skodvin and Steinar Andresen for helping us clarify this distinction.

4. These critiques are elaborated in greater detail in chapter 2 and in Betsill and Corell (2001).

5. The workshop was held at the Swedish Institute of International Affairs and funded by the Swedish Research Council and the US National Science Foundation (SES-0318165). The Workshop Report is available from the editors.

2

Analytical Framework: Assessing the Influence of NGO Diplomats

Elisabeth Corell and Michele M. Betsill

This chapter introduces the analytical framework that is at the core of this project and is the starting point for each of the empirical chapters in this volume. With this framework we mean to provide a basis for conducting systematic comparative analyses of NGO diplomats' influence on international environmental negotiations. We begin with a discussion of how to define NGO influence, particularly in light of the relationship between power and influence. This section culminates with a definition of NGO influence specific to our political arena—international negotiations on the environment.

We present our analytical framework in the second section. First we suggest ways of collecting and analyzing data related to NGO influence in a more systematic manner. Specifically, we identify two dimensions to NGO influence—participation in international negotiations and the subsequent effects on the behavior of other actors (e.g., states)—and encourage scholars to gather data on these two dimensions from a variety of sources. We suggest that researchers examine how NGO diplomats shape the negotiation process as well as outcome using process tracing and counterfactual analysis to analyze their data. The second part of our approach consists of a set of indicators that scholars can use to assess the overall influence of NGO diplomats in a particular negotiating context. Results may range from low levels of influence, where NGOs participate but have little effect on either the negotiation process or the outcome, to high levels of influence where NGO diplomacy is linked to effects on both process and outcome. Finally, we encourage analysts to identify the conditioning factors that enable or constrain NGO diplomats in their efforts to influence international environmental negotiations and to subject these factors to cross-case comparison.

Readers familiar with an earlier version of the framework (Betsill and Corell 2001) will note that the framework presented here has been revised. We made these revisions based on feedback from our contributors as they applied the framework to their respective cases and changes in our own thinking about NGO influence that came from ongoing conversations with scholars and NGO practitioners. We believe these revisions greatly enhance the utility of the framework in analyzing the influence of NGO diplomats in international environmental negotiations. Where appropriate, we highlight these changes and explain the reasoning behind them.

What Is NGO Influence?

In considering whether NGOs matter in global environmental politics, scholars seek answers to numerous questions. Do NGOs facilitate the evolution of a global civil society concerned with protecting the natural world? Do NGOs place issues on the international political agenda? Do they shape the outcome of international environmental negotiations? In each case the objective is to determine whether NGOs *influence* global environmental politics. It is surprising that few scholars define what they mean by NGO influence—the dependent variable of the studies they are undertaking. It is simply a discussion that is left out in most works. Two notable and commendable exceptions are Arts (1998) and Newell (2000).

The implications of failing to carefully define influence are threefold. First, without a clear understanding of what is meant by influence, analysts have little guidance as to what types of evidence should be collected. They often appear to be presenting evidence on an ad hoc basis and to have a bias toward evidence suggesting NGO diplomats made a difference in a given political process while ignoring evidence to the contrary. Second, the validity of claims of NGO influence can be challenged because there is no basis for assessing whether the evidence actually measures influence. Finally, it becomes difficult to compare NGO influence across cases because analysts rely on different types of evidence that may measure different aspects of influence. We thus begin with a more careful discussion of what we mean by NGO influence in this project.

Power versus Influence

Although influence is a basic concept in political science, it is difficult to define, at least partly because it is intimately linked to another difficult to define core concept—power. Explanations of influence vary depending on how influence is perceived to relate to power and the context in which the influence is exercised.[1] Scholars of international relations most often discuss power in terms of *state* power: state A has power if it can make state B do something that B would not choose to do (Dahl 1957). For instance, Holsti (1988: 141) defines power as the "general capacity of a state to control the behavior" of other states. According to Scruton (1996: 432), power is the "ability to achieve whatever effect is desired, whether or not in the face of opposition." Similarly Nye (1990: 25–26) defines power as the ability to achieve desired outcomes. Typical indicators of state power include gross national product, population, military capability or prestige.

Defining influence—and determining its relationship to power—has proven a challenging task. Holsti (1988), for example, views influence as an aspect of power, or a means to an end, but does not define influence. Scruton (1996: 262) states that influence is a *form* of power, but distinct from control, coercion, force, and interference:

It involves affecting the conduct of another through giving reasons for action short of threats; such reasons may refer to his advantage, or to moral or benevolent considerations, but they must have weight for him, so as to affect his decision. The influenced agent, unlike the agent who is coerced, acts freely. He may choose to ignore those considerations which influence him, and he may himself exert control over the influencing power.

But given his definition of power, Scruton clouds the difference between influence and power by including the possibility for the influenced agent to exert control over the influencing agent. It seems hard to discuss one without the other, but difficult to define them both so that they do not appear to be the same.

Cox and Jacobson (1973) attempted to avoid this problem by distinguishing more clearly between power and influence. They define power as "the aggregate of political resources available to an actor" (Cox and Jacobson 1973: 4). Power thus refers to *capability*. Cox and Jacobson (1973: 3) define influence as the "modification of one actor's behavior by that of another." In contrast to power, which can be calculated for any actor at a particular point in time, influence is seen as an emergent property that derives from the *relationship between actors*. Importantly Cox and Jacobson argue that power may or may not be converted to influence in any given political process. In other words, power does not necessarily guarantee that an actor will exert influence in its interactions. The key then is to understand the conditions under which an actor's capabilities result in influence.

NGO Influence in International Environmental Negotiations

Historically discussions of power and influence in international relations have focused on states. States are seen to possess military, economic, and political resources (power) that they use to exert influence. There is, however, growing awareness that non-state actors also possess capabilities that can be used to shape international outcomes. Mathews (1997: 50) argues, "(n)ational governments are not simply losing autonomy in a globalizing economy. They are sharing powers—including political, social and security roles at the core of sovereignty—with businesses, with international organizations, and with a multitude of citizens groups...."

Like states, NGO diplomats have access to a number of resources that give them power in multilateral negotiations. Although they rarely possess significant military capabilities, some NGOs have considerable economic resources, particularly in the private sector. Some argue that it is not their economic resources per se that make business/industry actors powerful but their central position in national economies and the international political economy (Levy and Newell 2000; Newell 2000; Rowlands 2001). This seems to have been the case in Burgiel's discussion of industry groups in the biosafety negotiations (chapter 4). Alternatively, Chatterjee and Finger (1994) argue that business/industry has a privileged position in international environmental policy making simply because "money talks." In their contribution to this volume, Andresen and Skodvin contend that this may hold for environmental NGOs as well. They note that Greenpeace reaped substantial financial resources from their campaign to generate public concern on whaling, which they reportedly used to shift the balance of power within the International Whaling Commission.

For many scholars and practitioners, knowledge and information are a key source of power for NGOs in world politics (e.g., Keck and Sikkink 1998; Corell 1999a).[2] In international environmental negotiations NGO diplomats often use their specialized knowledge in the hope of modifying actions taken by state decision-makers and/or altering how they define their interests. Such knowledge is a particularly valuable resource as international environmental issues are highly complex, and decision-makers often turn to NGO diplomats for help in understanding the nature of the problems and the implications of various policy alternatives under consideration. Knowledge and information enhance NGOs' perceived legitimacy in negotiations and may open up opportunities for influence. Each of the cases in this volume highlights the importance of knowledge and information as a crucial resource for NGO diplomats in international environmental negotiations.

As noted above, the relationship between power (capabilities) and influence is not direct. For states and non-state actors alike, the question remains *how* capabilities are translated into influence. Holsti (1988) identifies six tactics that states can use to exercise influence: persuasion, the offer of rewards, the granting of rewards, the threat of punishment, the infliction of nonviolent punishment, and the use of force. We find that many of these tactics are also used by NGO diplomats to exert influence in international environmental negotiations. Persuasion is perhaps the most widely used; NGOs spend considerable time trying to influence talks by persuading government representatives, who have the formal power to make the decisions, to accept the non-state actors' perspective. NGO diplomacy may also involve more coercive measures, such as threats and/or infliction of nonviolent punishment against states seen to be uncooperative. For example, many NGOs use a strategy of "blaming and shaming" in the hope of getting support for their positions by publicizing actions that interfere with the negotiations and/or noncompliance with previous commitments. NGOs may also threaten to interfere with economic activities in uncooperative states through boycotts (environmental NGOs) or by withholding investment (business NGOs). We see examples of such coercive tactics in some of our cases. The use of force is generally not a viable option for states or non-state actors during international environmental negotiations.

For the purposes of this project, we argue that influence occurs *when one actor intentionally communicates to another so as to alter the latter's behavior from what would have occurred otherwise.* In the earlier version of the framework, we used Knocke's (1990) definition of influence, which emphasizes information as the primary means of exerting influence within political networks. After many discussions with project participants on this matter, we concluded that this definition was too narrow and ran the risk of conflating power and influence. Information is one of many resources that NGOs may draw upon in their efforts to influence international environmental negotiations. Moreover project participants agreed that a definition of influence should be separate from the tools (power) used to achieve that influence. In this project we seek to analyze the observable effects of NGO participation in international environmental negotiations, regardless of the resources used by NGOs to bring about those effects (determining the relevant resources should be one foci of the research using the framework). Communication is a broader term that better captures the range of resources that NGOs use to influence international negotiations. Whether that communication occurs at the international or domestic level, in the form of technical information, claims of legitimacy, or threats, is for us to determine in each of the cases. We continue to emphasize that our definition of influence is tightly linked to a particular political arena—international environmental negotiations—and that it should not be read as a definitive statement of NGO influence in all areas of political activity.

Analytical Framework

Our definition of influence serves as the basis for the analytical framework at the core of this project. Zürn (1998: 646) argues that "[a]lthough there is a lot of good evidence about the role of transnational networks in international governance, more rigid research strategies are needed to determine their influence more reliably and precisely." The approach we introduce here represents such a research strategy. The framework begins by offering guidance for gathering and analyzing data related to NGO influence in a systematic fashion. We then develop a set of qualitative indicators that can be used to differentiate between three levels of NGO influence. Finally, we encourage scholars to consider the conditioning factors that enable or constrain NGO diplomats and help explain variation in NGO influence across cases.

One of our biggest challenges has been to develop an approach that simultaneously explains individual cases and helps us draw general lessons across cases. As with any research endeavor, this necessarily requires trade-offs. Mitchell (2002: 59) argues, "Carefully designed case studies often generate compelling findings that fit the case studied quite well but usually do so by sacrificing the ability to map those findings convincingly to many, if any, other cases" (see also Mitchell and Bernauer 1998). He adds that quantitative approaches usually have the opposite problem; they identify findings that hold relatively well across cases but do not explain any single case very well. Our aim is to give scholars the opportunity to highlight the unique aspects of each case while also providing a foundation for drawing more general lessons across cases. As a compromise we have chosen the method of structured, focused comparison for our cross-case analysis (George and Bennett 2005). We do not want to sacrifice the rich details that come forth in qualitative case studies. At the same time we have attempted to identify a set of general questions related to our research objectives that can be asked of each case study (structure), and we asked our contributors to focus on particular aspects of their cases (focus). The framework encouraged all contributors to ask questions both about what NGOs did in a given negotiating context as well as the observed effects, in particular, focusing on issue framing, agenda-setting, the positions of key states, and procedural and substantive outcomes.

Our framework relies heavily on triangulation—the use of multiple data types, sources and methodologies to analyze NGO influence in international environmental negotiations. "Triangulation is supposed to support a finding by showing that independent measures of it agree with it or, at least, do not contradict it" (Miles and Huberman 1994: 66). Triangulation can also help correct for the likely introduction of researcher bias in the development of indicators for assessing NGO influence. By the time a researcher gets to the point of identifying a set of possible indicators of NGO influence in a particular case, that person has likely spent a great deal of time studying and/or participating in the negotiation process. There is a danger of only looking at instances where NGO diplomats successfully exerted influence and ignoring failures (Arts 1998). Through the use of triangulation, researchers can develop qualitative confidence intervals about their conclusions on the level of NGO influence in multilateral negotiations on the environment.

Gathering Data

As discussed above, claims of NGO influence in international environmental negotiations can be strengthened by being more systematic in collecting data. In reviewing the literature on NGO influence in international environmental negotiations, we found that much of the evidence presented only indirectly measures influence, leading to validity concerns. Most scholars tend to rely on evidence regarding NGO *activities* (e.g., lobbying, submitting information or draft decisions to negotiators on a particular position), their *access* to negotiations (e.g., number of NGOs attending negotiations and the rules of participation), and/or NGO *resources* (e.g., knowledge, financial and other assets, number of supporters and their particular role in negotiations). Collectively this tells us a great deal about how NGO diplomats participate in international environmental negotiations. However, it is important to remember that participation does not automatically translate into influence; thus overemphasizing data on what NGOs do gives us an incomplete picture. To get a more accurate measure of NGO influence, researchers must also consider whether their efforts to shape multilateral negotiations are successful. If NGO diplomacy truly results in influence in international environmental negotiations, then it should be possible to observe the effects of NGO activities independent of those activities (King, Keohane, and Verba 1994).[3]

Our definition of influence highlights two dimensions of NGO influence in international environmental negotiations: (1) how NGO diplomats communicate with other actors during a negotiating process, and (2) alterations in the behavior of those actors in response to that communication. Given this definition of influence, researchers must look for evidence related to how NGOs participate in a specific negotiating process as well as evidence related to the behavior of other actors in the negotiations to assess whether influence has occurred (table 2.1). Data regarding participation (e.g., activities, access to negotiations and resources) address the first dimension by demonstrating whether and how NGOs diplomats communicated with other actors and identifying the specific content of such communication. We suggest that analysts may get at the second dimension by evaluating NGOs' *goal attainment* (see also Arts 1998; Biliouri 1999; Keck and Sikkink 1998; Williams and Ford 1999). A comparison of goals with observed effects speaks both to what NGO diplomats were trying to do when they communicated with other actors and whether those actors responded by altering their behavior.

It is important to note that the goals of NGO diplomacy may focus on both the *outcome* of the negotiations—such as the text of an agreement—as well as the *process* of the negotiations—such as the agenda (see Arts 1998; Betsill 2000). Perhaps the most obvious evidence of NGO influence is a connection between the text of the final agreement and NGO goals. If NGO diplomats influenced the negotiations, it is logical to expect congruence between ideas communicated by NGOs during the negotiations and the ideas embedded in the text of an agreement. An agreement may contain specific text drafted by NGO diplomats or reflect a general principle or idea introduced by NGOs during the negotiations. We argue, however, that researchers should not solely rely on evidence focused on the outcome of international environmental negotiations as a way to identify NGO influence. One problem is that there is frequently a gap between what NGO diplomats publicly demand and what they privately hope to achieve. For example, environmental NGOs often promote extreme positions as a strategy for pushing state decision-makers in new directions or for distracting their attention. They may have no expectation that these positions will actually appear in the final text. Moreover we may also observe the effects of NGO diplomacy on the negotiating process. For example, ideas communicated by NGO diplomats may show up in individual country statements, whose issues are (or are not) on the agenda, in the terminology used to discuss the issues

Table 2.1
Strategies for gathering and analyzing data on NGO influence (cells contain examples of questions researchers might ask)

Triangulation by:	Intentional communication by NGOs/NGO participation	Behavior of other actors/goal attainment
Research task: Gather evidence of NGO influence along two dimensions		
Data type	*Activities:* How did NGOs communicate with other actors? *Access:* What opportunities did NGOs have to communicate with other actors? *Resources:* What sources of leverage did NGOs use in communicating with other actors?	*Outcome:* Does the final agreement contain text drafted by NGOs? Does the final agreement reflect NGO goals and principles? *Process:* Did negotiators discuss issues proposed by NGOs (or cease to discuss issues opposed by NGOs)? Did NGOs coin terms that became part of the negotiating jargon? Did NGOs shape the positions of key states?
Data source	*Primary texts* (e.g., draft decisions, country position statements, the final agreement, NGO lobbying materials) *Secondary texts* (e.g., ECO, *Earth Negotiations Bulletin*, media reports, press releases) *Interviews* (government delegates, observers, NGOs) Researcher *observations* during the negotiations	
Research task: Analyze evidence of NGO influence		
Methodology	*Process tracing* What were the causal mechanisms linking NGO participation in international environmental negotiations with their influence?	*Counterfactual analysis* What would have happened if NGOs had not participated in the negotiations?

under negotiation, and/or in the general way the environmental problem is framed. Ignoring the effects NGO diplomats can have on the negotiation process simplifies and overlooks instances of NGO influence.

Contributors to this volume collected data from a variety of sources. We encouraged them to use as many different sources as possible, recognizing that each has different biases and/or strengths (Miles and Huberman 1994). Examples of primary documents used in this volume include the final agreement text, drafts negotiated along the way toward the final version, the official reports of each negotiation session, country statements, and NGO lobbying materials. Our contributors also made use of secondary documents, such as *ECO*, a publication produced by environmental NGOs during negotiating sessions to make their positions known, the *Earth Negotiations Bulletin*, which contains detailed daily and summary reports from the negotiations, as well as media reports and press releases. Several of our contributors also interviewed participants in the negotiations. For the most part these were conducted for other research purposes prior to constructing the case studies for this volume. Ideally, to control for potential bias, researchers should interview several different types of participants, including NGO diplomats, national delegates, and other observers who participated in the negotiations. As a rule, NGOs can be expected to overstate their influence on negotiations, and delegates can be expected to understate NGO influence (Arts 1998). Observers (e.g., UN agency staff) can therefore function as a control group. Finally, several of our contributors relied on evidence obtained by participating in and/or observing international environmental negotiations.

The particular conditions prevailing in the arena of international environmental negotiations give rise to some challenges in collecting data on NGO goal attainment. For example, as a result of failed efforts NGOs may revise their goals during the process, so the question becomes which of the goals should be considered as obtained? In addition NGO diplomats involved in an international environmental negotiation may not be coordinated enough in the beginning of the negotiation process to share common goals, so then, can goals acquired over time be considered to be obtained and at what point can the diverse group of NGOs be considered to have developed shared goals? Furthermore there are numerous groups involved in an international environmental negotiation, so whose goals should be examined, each individual organization's or the goals of the collective? While we recognize the complexities involved in applying this approach to NGO diplomacy in international environmental negotiations, a complementary approach that combines evidence on NGO participation with evidence related to goal attainment provides a richer picture of NGO influence by looking at the ways NGO diplomats communicate with other actors in multilateral negotiations as well as the subsequent effects.

Linking Participation to Influence
Evidence suggesting a connection between NGO activities in a particular negotiating context and observed effects enhances the plausibility of claims that NGO diplomats exerted influence. Such a connection raises the possibility that NGOs had some role in bringing about that effect. However, the risk of confusing correlation with causation remains. If a particular wording in the agreement text corresponds to the views of NGO diplomats, it does not necessarily follow that they were responsible for getting that text inserted into the agreement. It could be the case that other actors involved in the negotiations were promoting similar views. Giugni (1999) notes a similar challenge in the study of social movement consequences and argues that the problem of causality can be addressed, at least in part, by making careful methodological choices. Specifically, there is a need to elaborate the causal link between NGO participation and observed effects and to rule out alternative explanations. We contend that claims of NGO influence can be strengthened through the use of process tracing and counterfactual analysis (see table 2.1).

The fundamental idea behind process tracing is "to assess causality by recording each element of the causal chain" (Zürn 1998: 640; see also George and Bennett 2005, 206).[4] In the specific case of NGOs in international environmental negotiations, process tracing requires building a logical chain of evidence linking communication from NGO diplomats with other actors, actors' response/nonresponse, and the effects/noneffects of that communication. The first step is to demonstrate that NGO diplomats did engage in intentional communication with other actors. For instance, did they make an effort to provide negotiators with information about the nature of the problem, particular proposals, and so forth? As Knocke (1990: 3) notes, "influence is possible only when communication occurs." Communication is a two-way process. We must also consider whether the targeted actors were actually aware that communication had occurred and if so, how they responded. For example, if state delegates are unaware of an NGO proposal and/or if they do not consider the proposal to be politically viable, this suggests NGO diplomats have not been very effective at communicating their position. Analysts must thus question whether influence has occurred, even if there is a correlation between NGO participation and an observed effect.

Process tracing can take many forms (George and Bennett 2005). In this volume our contributors construct detailed narratives, often organized around hypotheses specific to the case. Process tracing helps analysts make causal inferences in single case studies and strengthens claims of NGO influence in any given negotiating context. Moreover, by specifying the causal links between NGO diplomacy and observed effects, process tracing can help scholars identify the conditions under which NGOs exert influence. Scholars can also use process tracing to rule out alternative explanations by trying to construct causal chains connecting the activities of other actors to an observed effect (see also Giugni 1999). If such a link cannot be made, the claim of NGO influence is strengthened.

Researchers should also consider whether the process and outcome of a given set of negotiations might have been different in the absence of NGO diplomats through the use of counterfactual analysis (Biersteker 1995; Fearon 1998; Miles and Huberman 1994; and Tetlock and Belkin 1996). Counterfactual analysis is an "imaginative construct" that considers what *might* have happened if one examined variable were removed from the chain of events (Biersteker 1995: 318). If the negotiations would have been the same regardless of the efforts or presence of NGOs, then it is more likely that they had little or no influence. If the negotiations would have been different had NGO diplomats not been involved, then the claim that they were responsible for an observed effect would appear to be strengthened. As Jon Elster has noted, "To distinguish causation from correlation we may point out that the former warrants the statement that if the cause had not occurred, then the effect would not have occurred, whereas no such counterfactual is implied by the latter" (quoted in Biersteker 1995: 318). We recognize the myriad difficulties related to constructing counterfactuals and have taken a fairly casual approach to counterfactual analysis in this project (George and Bennett 2005). We have encouraged authors to use counterfactual reasoning as one component of a broader analysis (along with process tracing) to help rule out alternative explanations and strengthen claims of NGO influence.

Assessing NGO Influence

One of the goals of our analytical framework is to encourage scholars to collect and analyze data on NGO influence in a more systematic fashion, and we believe that doing so will produce more robust claims of influence. In addition we believe it is possible to develop a set of indicators that enables us to assess the influence of NGO diplomats more precisely and that such assessments can serve as a basis for comparison across cases. We are not, however, in favor of a quantitative measure of NGO influence. We believe that precise quantification is futile and would only create a false impression of measurability for a phenomenon that is highly complex and intangible. Instead of "measuring" influence, we suggest that the influence of NGO diplomats can be qualitatively "assessed" in terms of high, moderate, or low levels of influence, by combining different types of evidence of NGO influence as illustrated in table 2.2 (for a similar approach, see Arts 1998: 74–85).

This is another example of how the framework has been revised since its earlier publication (Betsill and Corell 2001). We originally proposed a list of seven indicators, four of which addressed the ways that NGOs participate in multilateral negotiations and three which considered the subsequent effects. As contributors began using these indicators to assess NGO influence in their respective cases, a number of problems became apparent. First, the participation indicators were heavily biased toward information provision and left out other types of strategies and resources used by NGOs in some of the cases. Second, there was no clear link between the two types of indicators. Third, we realized that assessments of NGO influence ultimately rely more heavily on the indicators relating to the effects of NGO influence so these needed to be given greater weight.

Finally, the original set of effects indicators did not fully capture the range of potential effects we are likely to see in international negotiations on the environment and sustainable development. Table 2.2 is not necessarily a set of "new" indicators of NGO influence; it still contains data on participation and effects but the data is presented in a different format in an attempt to address some of the limitations noted above.

Our framework now identifies five indicators that can be used to assess the overall level of NGO influence in a particular set of negotiations. These indicators rely on the data and analytical methods outlined in table 2.1 and cover the range of effects we might expect to observe if NGO diplomats influence international environmental negotiations. The first three indicators focus on the potential effects of NGO diplomats on the negotiation process. *Issue framing* refers to how the environmental problem was conceptualized prior to and/or during the negotiations. A frame is "an interpretive schemata that simplifies and condenses the 'world out there' by selectively punctuating and encoding objects, situations, events, experiences, and sequences of action within one's present or past environment" (Snow and Benford 1992: 137). By framing (or re-framing) environmental problems, NGO diplomats can highlight particular aspects of a problem such as the driving causes and/or who has the responsibility to act, thereby establishing the boundaries within which states must formulate their responses (Betsill 2002; Chatterjee and Finger 1994; Humphreys 2004; Jasanoff 1997; Keck and Sikkink 1998; Williams and Ford 1999). For example, the problem of biosafety could be framed as a health issue or a trade issue, with implications for the types of information desired by negotiators and alternatives likely to be considered. If NGOs have an effect on issue framing, we would expect to see a correlation between the frames used by NGOs and those used by negotiators in their statements and/or as reflected in the final agreement. Issue framing may occur prior to the negotiation phase of the policy process (as in the case of desertification negotiations) or frames may change over the course of negotiations (as in the whaling and forests cases).

Another potential effect of NGO diplomacy relates to *agenda setting*. We view agenda setting as both a specific phase of the policy process (prior to the negotiation phase) and an ongoing process that occurs during the negotiation phase. Many scholars have found that NGOs

Table 2.2
Indicators of NGO influence (cells contain examples of the types of evidence analysts should include in the table and/or accompanying narrative)

Influence indicator	Evidence		NGO influence? (yes/no)
	Behavior of other actors . . .	. . . as caused by NGO communication	
Influence on negotiating process			
Issue framing	• How was the issue understood prior to the start of the negotiations? • Was there a shift in how the issue was understood once the negotiations were underway?	• What did NGOs do to bring about this understanding?	
Agenda setting	• How did the issue first come to the attention of the international community? • What specific items were placed on or taken off the negotiating agenda? • What were the terms of debate for specific agenda items?	• What did NGOs do to shape the agenda?	
Positions of key actors	• What was the initial position of key actors? • Did key actors change their position during the negotiations?	• What did NGOs do to shape the position of key actors?	

| Influence on negotiating outcome | Final agreement/ procedural issues | • Does the agreement create new institutions to facilitate NGO participation in future decision making processes?

• Does the agreement acknowledge the role of NGOs in implementation? | • What did NGOs do to promote these procedural changes? |
| | Final agreement/ substantive issues | • Does the agreement reflect the NGO position about what should be done on the issue? | • What did NGOs do to promote these substantive issues? |

catalyze international action by identifying to an environmental harm and calling upon states to do something about it (Charnovitz 1997; Newell 2000; Raustiala 2001; Yamin 2001; Gemmill and Bamidele-Izu 2002). Although our primary concern is on NGO influence during the negotiation phase of the policy process, we recognize that NGOs can open up opportunities for influence by drawing attention to a problem in the first place. We therefore encourage scholars to consider whether there is a link between NGO activities and how a particular problem came to the attention of the international community prior to the negotiation phase. At the same time the negotiation phase typically begins with setting up a framework for negotiation, which involves identifying the specific items to be addressed. For example, the Kyoto Protocol negotiations on climate change began with debates over the inclusion of developing country commitments and emissions trading. We therefore suggest that scholars consider whether NGO diplomats succeeded in placing issues on (or keeping issues off) the negotiating agenda.

Finally, we may observe the effects of NGO diplomacy in the *positions of key states* during the negotiations. Since state delegates ultimately decide on the text of an agreement, shaping the position of a key state or group of states can be an effective mechanism for NGO influence. Scholars may consider whether the initial positions of key states have been shaped by NGO diplomats. Moreover there may be evidence that a key state changed its position during the negotiations as the result of NGO activities. Andresen and Skodvin's chapter on whaling highlights the fact that NGO diplomacy aimed at shaping the position of key states may occur in the domestic context (e.g., by conducting public awareness campaigns or participating in domestic discussions) as well as in the international context (e.g., by lobbying state delegates at the negotiations).

The remaining two indicators consider the effects of NGOs on the final agreement, distinguishing between procedural and substantive issues. *Procedural issues* address how decisions are to be made in the future. NGO diplomats often wish to enhance opportunities for participation in future decisions by creating new institutions (e.g., advisory boards) and/or securing a role in implementation. NGOs may also shape the final text on *substantive issues* that make specific demands on member states. NGO diplomats typically have strong positions on what should be done to address an environmental problem, and these positions may be reflected in the agreement. In some cases we may find evidence that specific text proposed by NGOs appears in the agreement. More likely we may find elements of an NGO proposal and/or ideas consistent with NGO positions.

For each indicator, analysts should explicitly link evidence on what NGO diplomats did during the negotiations (i.e., how they participated) to evidence on how other actors behaved (subsequent effects). Determinations of NGO influence on any particular indicator require that analysts be able to provide specific evidence on both dimensions of influence (showing correlation) *and* that the data be analyzed using process tracing and/or counterfactual analysis to elaborate the causal link between NGO participation and observed effects (showing causality). Individually, no single indicator can point to a specific level of influence, but when aggregated, the indicators enable us to distinguish between high, moderate or low levels of NGO influence (table 2.3). In instances of low influence, NGO diplomats participate in negotiations but without effect. In other words, we find no evidence of NGO influence on any of the five indicators. Moderate influence occurs when NGOs participate and have some success in shaping the negotiating process. In these cases, we observe NGO influence on issue framing, agenda-setting and/or the positions of key actors (NGO diplomats need not influence each element of the process). The critical distinction between moderate and high levels of NGO influence relates to effects on the outcome of the negotiations. When NGO diplomacy can be linked to specific effects on the agreement text, NGOs can be said to have exerted a high level of influence in a particular set of negotiations.

Conditioning Factors

Finally, we encourage scholars to consider the factors that facilitate and/or constrain NGO diplomats in their efforts to influence international environmental negotiations. In this project we used an inductive approach to identify eight factors for the cross-case analysis: (1) NGO coordination, (2) rules of access, (3) stage of the negotiations, (4) political stakes, (5) institutional overlap, (6) competition from other NGOs, (7) alliances with key states, and (8) level of contention (see chapter 8). These factors

Table 2.3
Determining levels of NGO influence

	Low	Moderate	High
Description	• NGOs participate in the negotiations but without effect on either process or outcome.	• NGOs participate and have some success in shaping the negotiating process but not the outcome.	• NGOs participate in the negotiations and have some success in shaping the negotiating process. • NGOs' effects of participation can be linked to outcome.
Evidence	• NGOs engage in activities aimed at influencing the negotiations. • NGOs do not score a yes on any of the influence indicators.	• NGOs engage in activities aimed at influencing the negotiations. • NGOs score a yes on some or all of the process indicators. • NGOs score a no on all of the outcome indicators.	• NGOs engage in activities aimed at influencing the negotiations. • NGOs score a yes on some or all of the process indicators. • NGOs score a yes on one or both of the outcome indicators.

were derived from our contributors' detailed understandings of their respective cases as well as their general knowledge of the literature on NGOs in international environmental negotiations.

As discussed in chapter 1, our opportunistic approach to case selection precluded us from "testing" the explanatory value of any factor, since we made no determination about the appropriateness of the cases at the outset. However, as discussed in chapter 8, our analysis did identify a number of findings warranting future research. This should not be seen as an exhaustive list of all possible factors that condition NGO influence; the general literature on NGOs in international environmental negotiations suggests many others that could be examined systematically based on a more careful approach to case selection. In the following discussion, we review this literature in order to put our discussion of conditioning factors into context and to identify additional factors that could be analyzed using our framework.

Analysts frequently distinguish between those factors that emphasize the behavior or characteristics of NGOs (agency) and those that highlight the importance of context (structure) in explaining variation in NGO influence across cases. Most scholars combine elements of both structure and agency in their explanations of NGO influence in international environmental negotiations, and the distinction between agent-based and structural conditioning factors should not be overstated since they are often connected. Where structural factors are recognized, NGO diplomats may be able to act so as to take advantage of potential openings and/or avoid obstacles. Moreover, through their actions, NGOs may be able to change structural factors and open up new opportunities for influence.

Agent-based conditioning factors suggest that NGOs diplomats control their own destiny and can enhance their influence by adopting particular strategies and/or accumulating resources. For example, Dodds (2001) points to the importance of professionalization, arguing that NGO diplomats familiar with the technical language and procedures of multilateral negotiations are more likely to be successful in influencing the negotiations. Similarly many scholars stress that direct/insider tactics (e.g., lobbying) are more effective in the negotiation context than indirect/outsider tactics (e.g., protest; see Kakabadse and Burns 1994; Newell 2000; Gereffi, Garcia-Johnson, and Sasser 2001). Coordination among non-state actors is also seen to enhance their influence by amplifying their voice and promoting greater efficiency in gathering and disseminating information (e.g., Biliouri 1999; Chatterjee and Finger 1994; Keck and Sikkink 1998; Corell and Betsill 2001; Dodds 2001; Duwe 2001; Betsill 2002). Finally, some scholars contend that NGO influence is positively related to the possession of key resources, such as money and expertise (Chatterjee and Finger 1994; Kakabadse and Burns 1994; Biliouri 1999).

Alternatively, structural factors imply that NGOs are enabled or constrained by elements of the negotiating context. These factors help explain why, despite employing similar strategies or exhibiting similar characteristics, NGOs may have different levels of influence across cases. One set of structural factors underscores the institutional setting, or what social movement scholars refer to as the *political opportunity structure.* While there is considerable variation in how scholars define and operationalize political opportunity structure, McAdam (1996) finds that most emphasize the formal organizational/legal structure and power relations of a political system at a given time. There is some debate about whether this concept, which has been developed in the domestic context, travels to the international arena (see McAdam 1996; Kay 2005). However, we agree with Khagram, Riker, and Sikkink (2002) that international institutions have identifiable political opportunity structures and contend that the ability of NGOs to influence international environmental negotiations may be shaped by both aspects of the formal organizational structure in which the negotiation takes place and power relations among participating actors. Rather than construct a single measure of political opportunity structure, we find it more useful to think of political opportunity structures as clusters of variables and to analyze whether and how specific aspects of the institutional context shape NGO opportunities for influence (see Gamson and Meyer 1996).

In the context of international environmental negotiations, many scholars point to the rules for NGO access as an element of the organizational structure likely to constrain NGO diplomats. Where rules for access are restrictive, NGOs may be less likely to exert high levels of influence, since they have fewer opportunities for direct interaction with state delegates as well as limited access to information related to the negotiations (Corell and Betsill 2001; Raustiala 1997; Williams and Ford 1999). Moreover access rules may change as negotiations move from a general discussion to bargaining over specific text. In the latter stages of negotiations there may be less political space available to NGOs, since talks are more likely to be held behind closed doors with fewer participants in the room out of practical necessity. Finally, the legal nature of the negotiations may affect opportunities for NGO influence. NGO diplomats may be more influential in negotiations of framework agreements or nonbinding declarations where the political stakes are relatively low, since such agreements tend to articulate general principles and require few behavioral changes from states. In terms of power relations, opportunities for NGO influence may be constrained where there are significant cleavages between states (e.g., North–South conflicts) and/or other non-state actors promoting a conflicting agenda but enhanced by the availability of states allies (Arts 2001; Corell and Betsill 2001; Gulbrandsen and Andresen 2004).

Another structural factor that is more cultural than institutional relates to the way issues under negotiation are framed. Frames may enable or constrain NGO diplomats by creating a demand for particular types of information, thereby privileging some actors and limiting which proposals delegates consider seriously. For example, Corell and Betsill (2001) contend it is difficult for environmental NGOs to exert influence when environmental problems are linked to economic concerns because decision makers are more likely to focus on short-term economic costs than longer-term environmental costs. Similarly Williams and Ford (1999) found that the prevailing discourse of free trade within the World Trade Organization limited the political space available for environmental NGOs to promote their concerns about the environmental consequences of trade.

Conclusion

As we discussed in the introductory chapter, this volume has two central objectives. First, we seek to develop methodologies for strengthening claims of NGO influence in international environmental negotiations. The analytical framework we present here contributes to this objective by encouraging analysts to collect and analyze data on the influence of NGO diplomats in a more systematic manner. Moreover we argue that this systematic approach can be used to make more nuanced, qualitative assessments of NGO influence, which in turn allow for comparison across cases. The ability to compare across cases of NGO influence is essential to achieving our second objective: identifying a set of factors that condition the ability of NGO diplomats to influence international environmental negotiations. Such analysis is necessary to advance our theoretical understanding of the role of NGOs in global environmental politics by moving beyond the question of whether NGOs matter to examining under what conditions they matter.

Notes

1. This discussion draws heavily on Corell (1999a: 101–106).

2. We regard information as a set of data that have not been placed in a larger context. When information is placed within such a context, by relating it to previously gained knowledge, it becomes knowledge and can be used at a general level as the basis for assessments and action (Corell 1999a: 22).

3. For examples where this is done, see Arts (1998) and Newell (2000). Unfortunately, these represent the exception rather than the rule.

4. For examples of process tracing in the NGO literature, see Arts (1998), Close (1998), and Short (1999).

3

Environmental NGOs and the Kyoto Protocol Negotiations: 1995 to 1997

Michele M. Betsill

This chapter evaluates the influence of environmental NGOs (ENGOs) in the first phase of the negotiations of the Kyoto Protocol to the United Nations Framework Convention on Climate Change (UNFCCC), from August 1995 to December 1997.[1] The Kyoto Protocol was agreed upon in Kyoto, Japan, in December 1997, at the third Conference of the Parties (COP-3) to the UNFCCC. The Protocol responded to concerns that the commitments contained in the 1992 UNFCCC, which required industrialized countries to stabilize their greenhouse gas (GHG) emissions 1990 levels by 2000, were insufficient to meet its long-term objective of stabilizing atmospheric concentrations of GHGs. Debate focused on whether all countries (industrialized and developing) ought to be obligated to limit their GHG emissions and the extent to which those emissions should be limited.

This chapter uses the framework introduced in chapter 2 to assess ENGO influence during the negotiations, drawing on evidence collected during the period 1997 to 1999 through participant observation, interviews, and archival research (see Betsill 2000).[2] The chapter begins with a brief background on the international politics of global climate change, with particular attention to the first phase of Kyoto Protocol negotiations between 1995 and 1997. The next section presents evidence on ENGO participation in the negotiations, outlining their activities, access to the negotiations and resources. I then assess the level of ENGO influence on the negotiations, focusing on the issues of targets and timetables, emissions trading, and sinks. I find that while ENGO positions are not reflected in the Protocol's text, the environmental community did shape the negotiating process in a number of ways and thus had moderate influence. The final section considers the factors that shaped the ability of ENGOs to influence the negotiations.

International Politics of Global Climate Change

The international response to the threat of global climate change has centered on the negotiation of two multilateral agreements: the 1992 UNFCCC and its 1997 Kyoto Protocol. The UNFCCC was signed by more than 150 countries at the United Nations Conference on Environment and Development in Rio de Janeiro in June 1992 (see Mintzer and Leonard 1994; Paterson 1996; Betsill 2004).[3] As a "framework" convention, the UNFCCC created the basic architecture within which international efforts to address the threat of global climate change would take place. Specifically, the Convention established the ultimate objective of the international climate change regime as "stabilization of greenhouse gas concentrations in the atmosphere at a level that would prevent dangerous anthropogenic interference with the climate system" (United Nations 1992e, Article 2). In addition the UNFCCC obliged industrialized countries to aim to stabilize their GHG emissions at 1990 levels by 2000.

During the UNFCCC negotiations there was a great deal of debate about the nature of industrialized country commitments to control GHG emissions, with some participants (EU member states, small-island states, and ENGOs) arguing for *binding* targets and timetables for *reducing* emissions. The United States, however, with support from Japan, Canada, Australia, and New Zealand, called for *voluntary* commitments for *stabilizing* emissions without any clear timetable. The United States rejected binding targets and timetables on the ground that they were premature given remaining uncertainties about whether humans were causing climate change[4] and that reducing emissions could be devastating to the US economy (US Government 1991). Eventually negotiators gave in to the US demands. Given that the United States was responsible for more than one-quarter of 1990 global GHG emissions and fears of being placed at a competitive disadvantage, participants reasoned that it was essential to keep the United States engaged in the process of developing an international response to climate change, even if that meant adopted a weaker target (Nitze 1994; Paterson 1996).

At the first Conference of the Parties (COP-1), held in Berlin, Germany, the majority of participants agreed that the commitments contained in the Convention were insufficient to meet its long-term objective. The Conference adopted the "Berlin Mandate," which required Parties to negotiate a protocol by 1997 containing quantified and binding targets for reducing GHG emissions beyond 2000 (Conference of the Parties 1995). In addition the Berlin Mandate stated that the protocol would not contain new commitments for developing countries. The Ad Hoc Group on the Berlin Mandate met nine times between August 1995 and December 1997. Protocol negotiations also took place at COP-2 (Geneva) and COP-3 (Kyoto) of the UNFCCC (see Grubb, Vrulijk, and Brack 1999; Oberthür and Ott 1999; Betsill 2004). The negotiations were extremely complex, and ultimately many of the hard decisions were deferred until COP-3. It was only in the final hours of an unscheduled extra day of the Kyoto meeting that Parties were able to reach agreement on the text of the Kyoto Protocol.

During this phase of the Kyoto Protocol negotiations the debate centered on three issues: (1) who should be obliged to reduce greenhouse gas emissions, and in particular, what the role of developing countries should be; (2) how much should emissions be reduced by industrialized countries; and (3) how could such reductions be achieved.[5] The Protocol requires industrialized countries to reduce their aggregate GHG emissions 5.2 percent below 1990 levels by 2008 through 2012 (United Nations 1997a). These commitments are differentiated among Parties; each country has an individual target ranging from 8 percent reductions for EU member states to a 10 percent increase over 1990 levels for Australia and Iceland. The Protocol also allows Parties to use several "flexibility mechanisms" such as emissions trading to achieve those commitments cost-effectively. However, it left unresolved the specific rules and operational details for how such mechanisms could be used.

Because Parties with commitments were unlikely to ratify the Kyoto Protocol until they knew the "rules of the game," a second phase of Kyoto Protocol negotiations began almost immediately, culminating with the 2001 "Bonn Agreement" and "Marrakesh Accords." These agreements finalized the rules for implementation of the Kyoto Protocol and made it possible for industrialized countries to begin the process of ratification. The Kyoto Protocol entered into force on 16 February 2005, and as of April 2006 it has been ratified by 163 states (UNFCCC Secretariat 2006).[6]

ENGOs and the Kyoto Protocol Negotiations

ENGOs were extremely active participants in the Kyoto Protocol negotiations. More than forty organizations sent representatives to at least two of the negotiating sessions, with the largest delegations coming from Greenpeace, Friends of the Earth, and the World Wide Fund for Nature.[7] The environmental community was dominated by northern NGOs. Only one-fourth of the ENGOs came from the South, and these organizations typically sent only one or two representatives to the negotiations. The climate change secretariat provided some funds (raised from individual countries) for NGO participation; however, the funds were often insufficient.

ENGOs coordinated their participation in the Kyoto Protocol negotiations under the umbrella of the Climate Action Network (CAN). CAN was formed in 1989 for environmental organizations interested in the problem of climate change and today has more than 280 members (Climate Action Network 2003). CAN is a loose organization divided into eight regions, each with its own coordinator: Africa, Australia, Central and Eastern Europe, Europe/United Kingdom, Latin America, South Asia, Southeast Asia, and the United States/Canada. CAN served as the voice of the environmental community during the Kyoto Protocol negotiations. Members met daily during each negotiating session, and these meetings were an important forum for sharing information, debating issues, and coordinating lobbying efforts. In between negotiating sessions, some CAN members met regularly with other members in their respective regions (e.g., Europe) to devise strategies for lobbying particular governments.

During the period 1995 to 1997 CAN had four objectives.[8] First, CAN argued that the Protocol should include commitments for industrialized countries to reduce their GHG emissions 20 percent below 1990 levels by 2005. Second, they argued for strong review and compliance mechanisms to enhance the implementation of the commitments contained in the Protocol. Third, ENGOs objected to proposals to allow industrialized Parties to meet their commitments through emissions trading. Finally, CAN also opposed the idea of permitting Parties to get credit for emissions absorbed by sinks. The latter two objectives reflected CAN's position that industrialized states should achieve the majority of their emissions reductions through domestic policy changes. Throughout the negotiations CAN members framed the problem of climate change as an environmental crisis requiring immediate action (Betsill 2000).

CAN members employed a variety of strategies for promoting their position during the negotiations. Perhaps their most visible activity was the publication of a daily newsletter, *ECO*, at each of the negotiating sessions. *ECO*, which was widely read by all participants to the negotiations, served two purposes. First, it was a useful way for delegates to keep up with the day-to-day progress of the talks. Second, and most important in terms of exerting influence, CAN used *ECO* as a political forum for promoting their positions on a variety of issues, to discredit arguments put forth by opponents of emissions reductions (e.g., the oil-producing states and the fossil-fuel industry), and to put pressure on delegations to take aggressive measures to mitigate global climate change. Each issue contained a "fossil of the day" award given to the country that had most obstructed the negotiations the previous day. In addition CAN members used the pages of *ECO* to highlight their framing of climate change as an environmental crisis, regularly pointing to potential impacts such as more intense heat waves in Shanghai, stress to the Rocky Mountain ecosystem in the United States, damage to the Polish economy from more frequent floods, and significant declines in agricultural productivity in Africa and Asia (*ECO*, 7 March 1996; 5 March 1997; 4 August 1997; 6 August 1997).

CAN members also provided technical information to delegates. They publicized the potentially devastating impacts of climate change and conducted research on other scientific issues, such as the capacity of forests to serve as sinks. In addition several ENGOs produced their own cost–benefit analyses of various mitigation strategies and critiqued analyses produced by other organizations, highlighting how different assumptions lead to different predictions (Bernow et al. 1997; World Resources Institute 1997). During formal negotiating sessions, ENGOs held a variety of "side events" on technical issues related to the negotiations, although it should be noted that these events primarily attracted other NGOs and journalists rather than state delegates. CAN members devoted considerable time to evaluating proposals and identifying potential loopholes in the draft negotiating texts. As the negotiations progressed, such specialized knowledge was in demand by delegates who had to choose among policy options. It is important to note, however, that ENGOs did not have a monopoly on this type of knowledge and information during this period. Members of the scientific and business communities were also providing information on the physical impacts of climate change and the potential economic effects of various mitigation and adaptation options. These actors often provided contradictory information making it difficult for policy makers to uncover the "truth."

ENGOs had limited access to delegates during the negotiations, much more so than had been the case during the UNFCCC negotiations. This reportedly stemmed from an incident at a negotiating session prior to COP-1 where UN officials accused a prominent fossil-fuel lobbyist of orchestrating the floor debate by sending notes to OPEC delegates.[9] As a result NGOs were denied access to the floor during plenary sessions, and by the sixth negotiating session, delegates met primarily in closed-door, "nongroup" sessions from which NGOs were excluded altogether. Formally, NGOs were kept up-to-date through daily briefings with the Chair of the negotiations, as well as their respective state delegations. Informally, CAN members relied on the relationships they had developed with members of state delegations over the years, gathering information through corridor meetings and cell phone conversations. The use of cell phones was one particularly notable innovation during the Kyoto Protocol negotiations. On several occasions government delegates reportedly called environmental representatives to get their opinion on proposals being discussed in closed-door sessions, which enabled ENGOs to contribute to debates while not physically in the room (Boulton and Hutton 1997a).[10]

In addition CAN members resorted to more "subversive" measures; they lurked in corridors, hotel lobbies, and restrooms hoping to overhear conversations and/or corner key delegates; they even searched for draft documents and memos in trashcans and copiers. Overall, the problem of access was not insurmountable for the environmental community; as one representative noted, it just "wastes our time."[11] CAN members had to devote considerable time and resources to following the negotiations. Nevertheless, they continued to keep up to date on the status of the talks and were often able to prepare strategies to counter proposals before they were formally introduced.[12]

CAN members did have a few opportunities to participate directly in the Kyoto Protocol negotiations through informal roundtables and workshops organized to debate specific issues and proposals as well as formal statements delivered during plenary sessions. For example, during COP-2, Kiliparti Ramakrishna of the Woods Hole Research Center chaired a roundtable on possible impacts of industrialized emissions reductions on developing countries. Noting the involvement of the NGO community in this roundtable, Ramakrishna stated, "I hope delegates will agree with me that the inclusion of panelists from the nongovernmental community helped to enrich and enliven the discussion" (Ad Hoc Group on the Berlin Mandate 1996: 17). CAN representatives (like all NGOs) were permitted to deliver a formal statement to the plenary during each of the negotiating sessions, usually one statement by a representative of a northern ENGO and one from a representative of a southern ENGO. CAN used this platform to highlight the latest scientific information on climate change impacts, as well as the potentially negative economic impacts on developing countries if industrialized states failed to limit their GHG emissions.

While specialized knowledge was the primary source of leverage employed by CAN during the negotiations, there is some evidence that ENGOs also capitalized on their perceived role as shapers of public views about climate change and the appropriateness of governments' responses. Several governments complained about how they were portrayed by CAN. For example, at the second negotiating session, both the Philippines and the Netherlands objected that their positions on targets had been misrepresented in *ECO* (*ECO*, 1 November 1995; 3 November 1995). Some environmental groups also organized demonstrations and protest activities to draw public and media attention to the negotiations and the issue of climate change, although these were largely done on an individual basis rather than through CAN (Betsill 2000).

Assessing ENGO Influence

In the Kyoto Protocol negotiations ENGOs were active participants in that at each of the negotiating sessions they provided a great deal of written and verbal information to the negotiators. Although their ability to interact directly with the delegates was somewhat compromised, the problem of access was not insurmountable. These factors are only part of the story in assessing NGO influence in international environmental negotiations. This section examines whether these activities had any effect on the negotiation process and/or outcome. Specifically, I examine the negotiations around three of CAN's four objective areas: targets and timetables, emissions trading, and sinks. During this phase of the negotiations there essentially was no discussion on CAN's fourth objective, that the Protocol contain strong monitoring and compliance mechanisms. This issue was not taken up until the subsequent phase of Kyoto Protocol negotiations on implementation. In each of the issues examined, I find that while CAN's position was not reflected in the final text of the Protocol, the environmental community did shape the negotiating process both directly and indirectly.

Targets and Timetables

As mentioned above, the core issue in climate change negotiations between 1995 and 1997 was the establishment of binding targets and timetables for reducing GHG emissions. The central questions concerned *who* should be required to reduce their emissions and by *how much*? For many Parties the matter of developing country commitments had been settled at COP-1 and clarified in the Berlin Mandate. Critically, however, the call for developing country commitments remained a central part of the US negotiating position throughout this period, largely reflecting domestic politics and the power of the American fossil-fuel industry (Grubb, Vrulijk, and Brack 1999). Many developing country delegates, EU states as well as CAN members, accused the United States of violating the spirit of the UNFCCC and attempting to renegotiate the Berlin Mandate (Jaura 1997; Otinda, Ibrahima, and Sales Jr. 1997; Stevens 1997a). Moreover the call for developing country commitments was seen to ignore the principles of "common but differentiated" responsibilities and equity embedded in the UNFCCC.

The United States succeeded in inserting language on developing country commitments in the final draft negotiating text. In an all-night negotiating session on the final day of COP-3, Saudi Arabia, China, and India insisted that the text be deleted, while the United States, Russia, the Alliance of Small Island States (AOSIS), and Argentina argued that the language should remain. Ultimately the Chair of the negotiations made a unilateral decision to cut the text, referring back to the Berlin Mandate, and the United States did not press the issue further (Boulton and Hutton 1997; Brown and Leggett 1997). While this outcome reflected CAN's preference that the Protocol focus on industrialized country commitments, it is doubtful that CAN played a significant role in preventing developing country commitments from being included. Plenty of other actors had a clear self-interest in keeping such commitments out of the Protocol. The G-77 states and the EU most likely would have pursued this position even in the absence of the environmental community.

The negotiations over the specific levels of reductions were more complex. AOSIS, supported by CAN, put forth the first proposal calling for emissions reductions 20 percent below 1990 levels by 2005. Most industrialized states did not table their proposals until well into 1997, which meant that negotiations over the central issue were left until the very end. EU members, along with most developing countries, supported a two-step reduction target, calling for 7.5 percent reductions below 1990 levels by 2005 followed by a 15 percent reduction by 2010. The US position called for industrialized countries to stabilize emissions at 1990 levels within a five-year budget period (2008–2012). Critics argued that this position violated both the UNFCCC and the Berlin Mandate. Dr. Mark Mwandosya of Tanzania, Chairman of the G-77 stated, "It seems to me that the United States proposal is even less [than what was agreed upon in the UNFCCC]" (quoted in Stevens 1997a). Japan, Canada, Australia, and New Zealand proposed more modest reductions —between 3 and 5 percent below 1990 levels. Like the United States these countries faced domestic opposition from business and industry and thus shared concerns about the economic implications of addressing climate change.

In reality, the business/industry community was of three minds on the question of targets and timetables (Betsill 2000). A core group of fossil-fuel companies represented by the Global Climate Coalition (GCC) and the Climate Council opposed any international regulation. Between 1995 and 1997 a growing number of companies, including some former GCC members like British Petroleum, came to recognize climate change as a serious environmental threat as well as the greatest regulatory risk they had ever faced. Represented by groups such as the International Climate Change Partnership, these companies supported moderate emissions reduction targets, provided they allowed for sufficient flexibility to ensure cost-effectiveness. Finally, a number of "green" companies, including members of the solar and wind energy sectors, were in favor of international GHG regulations, foreseeing significant market opportunities should states be forced to move away from a dependence on fossil fuels. Such companies were represented by the American and European Business Councils for a Sustainable Energy Future.

The Protocol text requires that industrialized countries reduce their aggregate GHG emissions 5.2 percent below 1990 levels by the period 2008–2012, with each country committing to an individual target between an 8 percent decrease and a 10 percent increase (Article 3). This was largely a Japanese-brokered compromise between the American and EU positions, and by most accounts, a case of political horsetrading during the tough bargaining in closed-door sessions involving the EU leadership, the United States, and Japan over the final days (and ultimately hours) of COP-3. The targets are not based on scientific or economic analysis and are far below what the international scientific community says is necessary to stabilize atmospheric concentrations of GHGs.

The CAN/AOSIS proposal for 20 percent reductions was never seriously considered during the Kyoto Protocol negotiations because many delegates questioned its political feasibility. While CAN members framed the threat of global warming as an imminent environmental crisis requiring immediate action, this same sense of urgency was not reflected in the statements made by state delegates (Betsill 2000). Most states appeared to accept global warming as a legitimate environmental threat, though they did not sense that climate change was an impending crisis, noting uncertainty about the timing, magnitude, and distribution of climate change impacts. They were more concerned instead about how to mitigate the economic costs of controlling GHG emissions.

In the absence of CAN, the Kyoto Protocol targets might have been even weaker. Specifically, ENGOs appear to have played an important role in shaping the positions of the United States and the European Union, two key actors in the negotiations. An important turning point in the negotiations came with the decision of then–US Vice President Al Gore to attend the Kyoto meeting and to instruct the American delegation to be more flexible in its negotiating position. Several observers suggested that ENGOs were instrumental by generating media attention to the negotiations, which in turn may have increased the pressure for Vice-President Gore to attend the meeting.[13] One insider argued that the environmental community had nothing to do with Gore's decision to attend the meeting. According to this version of the story, Gore had always planned to attend but did not want to raise expectations in case something came up and he was unable to make the trip.[14]

Even if ENGOs did not influence Gore's decision to attend COP-3, they do appear to have influenced what he said once he arrived. The Vice President's speech included a last-minute addition (i.e., it was not included in the prepared text that was distributed before the speech) stating, "I am instructing our delegation right now to show increased negotiating flexibility if a comprehensive plan can be put in place..." (Gore 1997). Evidence suggests that American ENGOs convinced Gore to make this addition. Prior to his speech, the pages of *ECO* had been filled with calls for the United States to be more flexible in the negotiations, particularly in its opposition to a reduction target. High-level representatives of two American organizations reportedly conveyed this message to the Vice President (with whom they had established a close relationship during his tenure in the Senate) in a phone conversation during Gore's trip from the Osaka airport to the Kyoto convention hall.[15] Indeed, when Gore uttered the word "flexibility," two executives from one of these organizations smiled, shook hands and gave each other congratulatory pats on the back.[16] Following Gore's visit, the US delegation announced for the first time that it would agree to include targets for emissions reductions (rather than stabilization) in the Protocol.

In addition ENGO pressure seems to have been important in getting the European Union and developing countries to hold out for reduction targets before giving in on sinks and trading (Bettelli et al. 1997). By promoting an even higher reduction target, ENGOs made the EU proposal for 15 percent reductions look moderate. Moreover, Europeans are particularly concerned about how they are portrayed by the environmental community and thus were more willing to maintain a strong position than might otherwise have been the case. Commenting on the negotiations, EU Environment Commissioner Ritt Bjerregaard noted, "We are fortunate to have a lot of activist NGOs to push nations along."[17] Interestingly, many environmentalists expressed satisfaction (and sometimes shock) that the Protocol contained any reduction commitments at all.[18]

This analysis highlights the interaction between domestic and international channels of NGOs influence. At the domestic level, the environmental community failed to shape the US position, losing out to an aggressive campaign by members of the American fossil-fuel industry (see Betsill 2000). Groups like the GCC succeeded in framing the issue of climate change as a significant economic threat and mobilized opposition in Congress and the public, which in turn limited the ability of the Clinton administration to put forward a progressive position on targets and timetables. However, at the international level, the GCC did not have sufficient resources and organizational capabilities to ensure that the United States stuck to its position of opposing any reduction targets. Through CAN, American ENGOs joined their European counterparts in regular meetings with EU delegates, promoting their position that the Protocol must contain reduction targets and reminding European decision-makers that their constituents supported a commitment (thanks in large part to the domestic work of European ENGOs).[19] In turn, the EU states (along with the G-77) maintained pressure on the United States to accept reduction targets.

Emissions Trading
As noted above, negotiations over reduction targets were closely tied to debates about the use of market mechanisms, including emissions trad-

ing. The United States, along with Japan, Canada, Australia, and New Zealand and with support from many of the industry groups, argued for maximum flexibility and the use of market mechanisms to enable states to meet their commitments in a cost-effective manner (Ad Hoc Group on the Berlin Mandate 1997b). The European Union, most CAN members, and the majority of developing countries objected on the grounds that it would allow those industrialized countries (in particular, the United States) that had been responsible for the vast majority of greenhouse gas emissions to date to buy their way out of making changes in their consumption patterns at home, hence going against the "polluter pays" principle enshrined in the UNFCCC.

The issue of emissions trading also created tensions within the environmental community. Efforts to create an international GHG emissions trading regime were largely based on the US experience with sulfur dioxide emissions, where the Environmental Defense Fund (EDF—now Environmental Defense) and to a lesser extent the Natural Resources Defense Council were particularly instrumental. During the Kyoto Protocol negotiations, EDF advocated trading as a viable option for implementing the agreement's reduction targets and reportedly had a hand in drafting the language that appeared in the negotiating text.[20] However, many other CAN members objected to this position, arguing it would do little to bring about the fundamental social changes necessary to promote the sustainable use of resources and a just international economic order.[21] This was one of the few instances of in which divisions between members of the environmental community became apparent outside of CAN.

As noted above, trading language made its way into the draft negotiating text and was considered during the all-night review session in the final hours of COP-3. When China and India objected to the language, the Chair of the negotiations reminded delegates that certain industrialized Parties (the United States) required trading in order to accept any legally binding reduction commitments and asked delegates to reflect on the consequences of refusing to accept trading language (Bettelli et al. 1997). After three hours of debate on this one article, the Chair called for a brief recess and came back to say that trading language would stay in the Protocol (Article 17) but that specific details about how a trading regime might operate would be worked out later (Brown and Leggett 1997; Stevens 1997b; Grubb, Vrulijk, and Brack 1999).

While unsuccessful in their effort to keep trading off the agenda and out of the Protocol, CAN members seem to have been at least partially responsible for the Parties being unable to agree on the rules for trading during this phase of the negotiations. CAN slowed the debate by raising concerns about what came to be known as "hot air" emissions trading, which refers to the ability of a country whose emissions are below its legally binding limits to trade the difference (*ECO*, 27 October 1997; 28 October 1997). At the time the Kyoto targets were negotiated, emissions in Russia and Ukraine were more than 30 percent below 1990 levels due to economic decline, yet each country secured a stabilization commitment under the Protocol. As a result both Russia and Ukraine have a supply of surplus hot air emissions—emissions that exist only on paper—that can be sold to countries like Japan and the United States who might have difficulty meeting their targets through actual domestic reductions. Environmentalists argued:

If [Russia] received the same target as all of the other Annex I countries and this target was less than a 5–10% reduction in emissions from 1990 levels, the volume of "paper" carbon credits would be enough to eliminate the effect of the emissions reduction commitment (*ECO*, 27 October 1997).

The term "hot air" has since entered into the vernacular of climate change negotiations and continues to be a central part of discussions on emissions trading. Many participants and observers to the negotiations credit CAN with introducing the term hot air and placing this potential loophole in emissions trading on the negotiating agenda (Bettelli et al. 1997: 15).

Sinks

The emissions reduction targets formalized in the Kyoto Protocol are based on states' net emissions of GHGs—gross emissions minus emissions absorbed by "sinks." Sinks are "physical and biological processes...which remove carbon dioxide from the atmosphere" (UNEP and WMO nd: 13). Although many states allowed for removals by sinks in their emissions reductions proposals, focused negotiations on sinks did not begin until November 1997 (less than one month before

COP-3). The negotiations centered around New Zealand's "gross-net" approach whereby countries would define their 1990 baseline in terms of gross emissions (sources) while defining their allowed emissions for the budget period in terms of net emissions (sources minus sinks) (Depledge 2000). While most states generally supported the inclusion of sinks, the European Union, Japan, and many developing countries expressed concern about including them in the first commitment period, citing problems with accounting methodologies and other technical issues (Ad Hoc Group on the Berlin Mandate 1997a). CAN members also opposed the inclusion of sinks in the Protocol on the methodological ground that "[r]obust techniques for calculating the removal of greenhouse gases have not yet been prepared by the [Intergovernmental Panel on Climate Change]" (*ECO*, 30 November 1997). The United States, Canada, Australia, and New Zealand viewed sinks as another element of "flexibility" that was essential for achieving emissions reductions in a cost-effective manner (Oberthür and Ott 1999). Like the question of emissions trading, debates about sinks were also intimately linked to targets, and it proved impossible to establish targets without a clear understanding of how sinks would be treated in the Protocol. To this end, the contact group on sinks met almost around the clock during COP-3 (Depledge 2000). In the end the Protocol does allow for countries to get credit for emissions absorbed by sinks (Article 3.3). However, technical decisions about how sinks would be treated were left for future negotiations.

Once again, ENGOs were unable to keep an issue off the negotiating agenda and out of the Protocol. ENGOs could only shape how the sinks debate developed by raising concerns, and in their absence, the issue might have been less contentious. Michael Oppenheimer of the Environmental Defense Fund and Daniel Lashof of the Natural Resources Defense Council, both of whom participate in the Intergovernmental Panel on Climate Change, were the primary spokespersons for ENGOs on the issue of sinks, and they regularly met with delegates participating in the sinks working group.[22] Although the Protocol does permit countries to get credit for emissions absorbed by sinks, delegates were unable to agree on how these levels would be calculated. One delegate noted that ENGOs managed to raise distrust about sinks among some participants and that several delegations refused to even talk about sinks.[23] CAN's criticism also reportedly influenced France's position on sinks. Dominique Voynet, French Minister for Territorial Planning and the Environment, acknowledged that ENGOs had prompted her country to oppose the inclusion of sinks language in the Protocol (quoted in *ECO*, 9 December 1997).

Level of Influence

ENGOs had little effect on the *outcome* of the Kyoto Protocol negotiations during the period 1995 to 1997. CAN's positions are not reflected in the Protocol's text. While CAN members advocated for industrialized countries to reduce GHG emissions 20 percent below 1990 levels by 2005, the Protocol only requires an aggregate of 5.2 percent reductions below 1990 levels over the period 2008 to 2012. ENGOs opposed the inclusion of emissions trading and sinks, both of which appear in the final treaty text. Nevertheless, ENGOs played an important role behind the scenes, influencing the Kyoto Protocol negotiations in ways that cannot readily be observed by looking solely at the final text. In other words, CAN influenced the negotiation *process*.

By the framework presented in chapter 2, ENGOs can be said to have exerted a moderate level of influence on the Kyoto Protocol negotiations between 1995 and 1997 (table 3.1). ENGOs were active participants in the negotiations, and they had some success in shaping the negotiating process but not the outcome. While ENGOs did not convince delegates to frame climate change as an imminent environmental threat, they did affect the negotiating agenda by catalyzing debate on emissions trading and sinks. They also shaped the positions of key states on the issue of targets and timetables: the EU through domestic and international channels and the United States through international channels. However, they failed to get delegates to discuss compliance and review mechanisms, an issue central to the ENGO position.

Explaining ENGO Influence

During the Kyoto Protocol negotiations, a number of factors enabled and constrained ENGOs in their quest to influence the negotiating process and outcome. These factors can be divided into three broad categories: those related to the nature of the issue, those related to the institutional context in which the negotiations took place, and those related to the NGOs themselves.

Nature of the Issue

Scientific uncertainty about the timing, magnitude, and distribution of potential impacts of climate change hindered development of a consensus regarding the appropriate response. Environmentalists tended to focus on the most severe projections. They often attributed extreme weather events to global warming when members of the scientific community were unwilling to do so, thereby undermining the credibility of ENGO claims. The linkage between climate change and the global economy also limited ENGO impact during this period. Controlling GHG emissions is likely to have implications for global energy prices and industrial production, issues at the heart of industrialized economies. As a result some of the most powerful state (e.g., United States) and non-state (e.g., the oil industry) actors in the international system routinely emphasized the potential negative economic effects of global regulation (Ad Hoc Group on the Berlin Mandate 1997b; BNA 1997; Cushman Jr. and Sanger 1997; Knight 1997). In particular, the fossil-fuel industry mobilized considerable resources during this period to reinforce this point in its bid to stall the negotiations.

This high level of contention over economic issues limited the political space available to ENGOs, making it difficult for the environmental community to challenge the economic arguments against climate regulation. They had significantly fewer financial resources at their disposal than the fossil-fuel industry. ENGOs emphasized the long-term costs of *not* acting, while delegates were much more concerned with the short-term costs of controlling GHG emissions (Ad Hoc Group on the Berlin Mandate 1997b). CAN members did provide economic analyses suggesting that industrialized countries could reduce their emissions at minimal costs, but there were an equal number of studies indicating that international regulation would lead to economic ruin. Finally, ENGOs sought to make their case for aggressive action to combat climate change by linking current activities (e.g., emission of GHGs in industrialized

Table 3.1
ENGO Influence in the Kyoto Protocol negotiations

Influence indicator	Evidence		NGO influence? (yes/no)
	Behavior of other actors...	...as caused by NGO communication	
Influence on the process			
Issue framing	• Delegates viewed problem as a legitimate threat but not an imminent crisis. • Delegates showed concern about economic effects.	• NGOs viewed problem as an imminent environmental crisis.	No
Agenda setting (during the negotiation phase)			
• Emissions trading	• Delegates adopted the language of "hot air."	• NGOs called provisions allowing Russia to sell credits for emissions that didn't exist "hot air."	Yes
• Sinks	• Delegates were reluctant to rush to adopt to provisions allowing credit for emissions absorbed by sinks.	• When proposals for sinks were introduced, NGOs raised a number of concerns about measurement and equity.	Yes
Positions of key actors			
• United States	• Stabilization and flexibility	• Original position reflected domestic pressure from the fossil-fuel industry; position changed due to mobilization of outside pressure from ENGOs and connections with Gore.	Yes

• European Union	• Delegates held out on reduction targets before giving in to US on trading.	• Pressure came from ENGOs; states were concerned about how they were viewed at home.	Yes
Influence on negotiating outcome	Final agreement/ procedural issues		No
	Final agreement/ substantive issues	• Text only requires 5% reductions and allows for emissions trading and sinks.	• NGOs had pushed for 20% reductions and opposed the use of flexible mechanisms.	No
Level of influence			Moderate

countries) to effects that would be distant in both time and space (e.g., more severe droughts in Africa in the 2050). In contrast, the fossil-fuel industry was able to exploit linkages that were more short-term and direct.

The highly technical nature of the Kyoto Protocol negotiations created a demand for specialized knowledge on issues such as sinks and emissions trading. Whereas many of the delegates participating in the negotiations were new to the issue of climate change, a number of environmentalists had been engaged in climate change politics for nearly a decade. As a result the environmental community had developed technical expertise and was able to help delegates, especially those from the South, make sense of the myriad proposals about how to reduce GHG emissions (Newell 2000). In an extreme example, the London-based Foundation for International Environmental Law and Development advised members of the Alliance of Small Island States and represented some of the alliance's members in the negotiations. At the same time negotiators came to focus on techno-fixes as a solution to curbing GHG emissions. Given the link between climate change and the energy sector, delegates began looking to new energy technologies, a sphere in which industry organizations had greater leverage. In other words, as the negotiations turned toward the specific details of how to achieve emissions reductions, ENGOs became less central to solving the problem.

Institutional Factors
Although the rules governing NGO access to the Kyoto Protocol negotiations were somewhat restrictive, this does not seem to have been a significant factor shaping CAN's influence during the period 1995 to 1997. By that time many of the environmental representatives had been participating in political debates about climate change for nearly a decade. They had developed personal relationships with state delegates, so they were able to keep up to date on the status of the negotiations and to contribute to debates. According to one environmentalist, "Negotiators often come up to us to comment on issues and debate them with us, which stimulates thinking—and acting—in the right direction."[24] Moreover the introduction of cell phones into the negotiating venue made the lack of physical presence in meeting rooms less problematic.

The alliance between environmentalists and the Europeans during the Kyoto Protocol negotiations was of particular import. Both groups had been disappointed by the outcome of the UNFCCC negotiations and wished to push the United States to accept binding targets and timetables to reduce GHG emissions. The Europeans became champions of ENGO positions on trading (Bettelli et al., 13 December 1997). As noted earlier, this alliance with a key state actor was essential for environmentalists as it allowed them to exert pressure on the Europeans to hold out for reductions from the United States before giving in on emissions trading.

At the same time the institutional context in which the negotiations took place made it unlikely that the ENGO proposal for a 20 percent reduction target would be seriously considered as a viable outcome. The fact that the Protocol would be a formal international treaty with binding commitments raised the political stakes of the negotiations. When environmentalists first proposed the 20 percent reduction target at the 1988 Toronto Conference, they faced little resistance from delegates, since any standard adopted in the conference statement would not be legally binding (Betsill 2000). Once the negotiations shifted to the United Nations and the development of more formal international law, delegates had to balance their desire to move forward on the issue of climate change with the need to appease domestic constituencies. As the political stakes became higher, delegates were more likely to focus on those proposals that could be portrayed as moderate.

NGO Profile

Many observers point to the ability of ENGOs to coordinate their activities and speak with one voice as central to their influence on the Kyoto Protocol negotiations.[25] Without such coordination, statements made by any one group would have been viewed as representing narrow interests rather than the broader interests of the environmental community. It is important to note that environmentalists were not the only non-state actors engaged in the Kyoto Protocol negotiations. Groups representing the interests of business and industry often outnumbered environmental organizations. The presence of other non-state actors meant that ENGOs were competing for the attention of state decision-makers. Moreover CAN members had to devote time to keeping informed on the positions of business and industry organizations and challenge arguments that went against their own agenda.

During this period many ENGOs prioritized their international activities over their national and regional climate change campaigns. The ability of non-state actors to influence international negotiations depends on their ability to put pressure on all levels simultaneously, to operate through transnational, international and domestic channels (Keck and Sikkink 1998). In particular, it can be important to shape the negotiating positions of key states before they arrive at the international negotiations. CAN members appear to have recognized the importance of working at the national and regional levels as reflected in the division of the network into autonomous regional organizations. However, during the Kyoto Protocol negotiations, American ENGOs failed to put this principle into practice. While they spent countless hours (not to mention hundreds of thousands of dollars) attending international climate change meetings, their domestic activities were largely confined to lobbying the Clinton administration. Environmentalists virtually ignored Congress and the general public; a void quickly filled by representatives of the fossil-fuel industry who ultimately succeeded in framing the climate change issue in the United States. Of particular note, several members of the administration blamed the environmental community for the lack of public support for reduction targets.[26] In retrospect, some American ENGO representatives acknowledge that they erred in not focusing more directly on the domestic political arena in the United States during the Kyoto Protocol negotiations.[27] At the same time they were able to use their access to Vice President Gore, as well as their transnational and international connections, to move the negotiations closer to their preferred outcome.

Conclusion

This case demonstrates the necessity of differentiating between NGO *activities* and NGO *influence* in the realm of international environmental negotiations. While the environmental community, working together through CAN, was an extremely active participant in the Kyoto Protocol negotiations during the 1995 to 1997 period, merely "showing up" was not sufficient to ensure they achieved their goals. At the same time, evaluating NGO influence solely on the basis of negotiation *outcomes* misses a significant part of the story about the effects of NGOs on negotiation *processes*. CAN members directly shaped the nature of debates around emissions trading and sinks, and indirectly influenced negotiations on targets and timetables by putting pressure on delegates from the European Union and the United States. This case raises a number of questions about the how institutional rules, the nature of the issue, and the selection of particular strategies mediate between NGO activity and influence in international treaty negotiations. These questions could be addressed in future research.

Notes

I wish to thank Paul Wapner, Stacy VanDeveer, Felix Dodds, Tore Brænd and the participants at the Stockholm Workshop on NGO Influence in International Environmental Negotiations for their helpful comments on a previous draft.

1. Negotiations during this period focused on drafting the text of the Kyoto Protocol. Subsequent negotiations (1998–2001) focused on the rules for implementing the Protocol and other matters left unaddressed by the text. See Betsill (2004) for more details.

2. I attended COP-3 in December 1997; conducted semi-structured interviews with more than 30 environmentalists, government delegates, industry representatives, UN personnel, and scientists affiliated with the negotiations, and reviewed all official UN documents as well as numerous media and other secondary reports on the Kyoto Protocol negotiations. This research was supported by the Institute for the Study of World Politics and the University of Colorado-Boulder. For specific details on this data, see Betsill (2000) or contact the author directly.

3. On the influence of NGOs on these negotiations, see Arts (1998), Betsill (2000), and Newell (2000).

4. It was not until the second assessment report, released by the Intergovernmental Panel on Climate Change in 1995, that the scientific community made its now infamous statement that the "the balance of evidence suggests a discernible human influence on global climate" (IPCC 1995).

5. These issues are discussed in greater detail below.

6. The United States and Australia have formally announced that they do not intend to ratify the Protocol.

7. Participation lists available from the UNFCCC Secretariat in Bonn.

8. These positions were regularly articulated in the pages of *ECO*, the daily newsletter published by CAN members at each of the negotiating sessions. Available at ⟨http://www.climatenetwork.org/eco/backissues.html⟩.

9. Interviews, ENGO 9, 2/18/99; UN Representative 3, 11/11/97.

10. Interviews: ENGO 3, 12/8/97; ENGO 7, 2/16/99; ENGO 9, 2/18/99; Delegate 4, 2/22/99.

11. Interview, ENGO 4, 10/26/98.

12. Interviews, ENGO 4, 10/26/98; ENGO 8, 2/17/99; ENGO 9, 2/18/99; ENGO 10, 2/22/99; US Delegate 4, 2/22/99.

13. E-mail correspondence, ENGO 13, 4/14/99; Interviews, ENGO 7, 2/16/99; Delegate 1, 12/9/97; Delegate 5, 2/22/99.

14. Interview, Delegate 4, 2/22/99.

15. Interview, Scientist 1, 7/28/99.

16. Author's notes, 12/8/97.

17. Author's notes, 12/9/97.

18. E-mail correspondence, ENGO 11, 3/3/99; ENGO 13, 4/14/99. Interviews, ENGO 4, 10/26/98; ENGO 6, 2/16/99; ENGO 8, 2/17/99; ENGO 9, 2/18/99.

19. This is an example of what Keck and Sikkink (1998) refer to as the "boomerang."

20. Interview, ENGO 10, 12/9/97.

21. Interviews, ENGO 4, 10/26/98; ENGO 7, 2/16/97; ENGO 8, 2/17/99; ENGO 10, 2/22/99.

22. Interviews, ENGO 4, 10/26/98; ENGO 6, 2/16/99; ENGO 7, 2/16/99; ENGO 9, 2/18/99; Scientist 2, 12/11/98.

23. Interview, Delegate 4, 2/22/99.

24. Interview, ENGO 11, 3/3/99.

25. Interview Scientist 6, 3/19/99; Business NGO 5, 2/23/99; e-mail correspondence ENGO 11, 3/3/99.

26. Interviews, Delegate 2, 9/1/98; Delegate 3, 9/2/98; Delegate 4, 2/22/99.

27. Interviews, ENGO 7, 2/16/99; ENGO 9, 2/18/99; ENGO 10, 2/22/99.

4

Non-state Actors and the Cartagena Protocol on Biosafety

Stanley W. Burgiel

The presence of non-state actors within international policy circles has been steadily increasing in recent decades, particularly, in multilateral environmental negotiations addressing such issues as biodiversity, genetically modified organisms (GMOs), and climate change. The multifaceted nature and highly subjective elements involved in negotiating processes make it difficult to gauge the impact of these groups, whether promoting environmental advocacy or business interests. The present study seeks to analyze the influence of non-state actors in negotiations to develop the Cartagena Protocol on Biosafety under the Convention on Biological Diversity (CBD).

Adopted after more than five years of intense negotiation, the Protocol has established a transparent procedure for state decisions on whether to import certain GMOs based on their potential environmental and health impacts. Environmental nongovernmental organizations (ENGOs) were active throughout the process and supported incorporating provisions such as the precautionary principle, socioeconomic considerations, documentation requirements, and liability in the agreement. Industry representatives became more involved in the process as the implications for trade in GMOs became increasingly apparent. While ENGOs and industry followed the discussions, the concluding stages of the negotiations were dominated by divisive debates among governments over the trade and environmental ramifications of a number of controversial provisions. By the close of negotiations, five distinct blocs of states came to dominate the discussions.

This chapter will start by briefly reviewing the Cartagena Protocol on Biosafety, its negotiating history, and the major issues in the negotiations:

the agreement's scope, trade-related concerns, decision-making criteria, and the responsibilities of those exporting GMOs. This will serve as a foundation for examining the roles of ENGOs and industry groups and their abilities to influence the negotiations.

The analysis draws on data collected through the author's personal involvement in and attendance at the negotiations as a writer for the *Earth Negotiations Bulletin* from 1996 to 2000. References to ENGO and industry documents included herein are representative of a much larger set of position papers, statements and notes from discussions with non-state actors and industry contained in Burgiel (2002), which provides a more comprehensive view of the Protocol negotiations as a whole. Such discussions and in-depth review have been abbreviated for the present chapter's focus on the influence of non-state actors.

Cartagena Protocol on Biosafety

The Cartagena Protocol on Biosafety was established to promote the safe international transfer, handling, and use of living modified organisms (LMOs), through an advance informed agreement procedure for the import of LMOs destined for intentional introduction into the environment.[1] This Protocol targets genetically modified (GM) seeds, plant material, and nursery stock intended for planting in fields, gardens, forests, and elsewhere, without further regulation. (GM fish and animals could have been covered but at the time were not in such advanced stages of commercial development to be of immediate concern.) The procedure requires an exporter to inform a country of its intent to ship LMOs. The country of import can then undertake a risk assessment on any potential impacts and make a decision on whether to import the specific LMO. The importing country can also employ the precautionary approach if there is cause to believe that harm can occur despite insufficient scientific evidence.

The Protocol includes an alternative procedure for LMOs intended for direct use as food, feed, or processing (LMO-FFPs), and thereby not intended for planting or release into the wild. Such LMO-FFPs are the basic agricultural commodities that constitute the bulk of international trade in LMOs. The difference with the full procedure for intentional introductions is that a country developing an LMO-FFP announces its commercialization and then other countries are responsible for notifying the producing country about whether they wish to undertake a risk assessment and make a decision on whether to allow imports. If a country does not respond, that particular LMO-FFP can then be exported to that country without further restrictions. The Protocol also includes requirements for documentation and identification of shipments, liability and redress provisions, provisions on capacity building and financial resources particularly for developing countries, and a Biosafety Clearing-House to facilitate information exchange.

Negotiating History
The negotiations to develop a protocol on biosafety under the CBD extend back to the negotiation of the Convention from 1990 to 1992 and its inclusion of Article 19.3, which calls for consideration of the need for and modalities of a protocol on the safe transfer, handling, and use of LMOs that may have an adverse effect on biodiversity (United Nations 1992b). Debates on the need for a protocol spanned the two Intergovernmental Committee meetings on the CBD (October 1993 and June–July 1994) through to the First and Second Conferences of the Parties (November–December 1994 and November 1995). With agreement at the Second Conference of the Parties to move ahead with a protocol, the actual negotiations included six meetings of the Biosafety Working Group (BSWG), the failed Extraordinary Conference of the Parties (ExCOP) in Cartagena, three sets of informal consultations in Montreal and Vienna, and the final, "resumed" ExCOP, which adopted the agreement in Montreal (January 2000).[2]

By the close of negotiations, states had separated into five discrete negotiating blocs:

• The Miami Group (Argentina, Australia, Canada, Chile, the United States, and Uruguay)
• The Like-Minded Group (the majority of developing countries)
• The European Union
• The Compromise Group (Japan, Mexico, Norway, Singapore, South Korea, Switzerland and, during the resumed ExCOP, New Zealand)
• The Central and Eastern European bloc of countries (CEE)

From the start, most developed countries and industry stated that there was no need for such an instrument, noting the sufficiency of domestic legislation or voluntary guidelines and codes of conduct, such as those being developed by the United Nations Environment Programme. Developing countries, particularly Malaysia, Indonesia, and India, as well as ENGOs supported the negotiation of a binding agreement, noting that domestic legislation could not address the international dimensions of LMO transfers. They also argued that voluntary guidelines would be ineffective in committing the biotechnology industry to act responsibly, based on past experiences with the chemical industry and the increasing number of undisclosed field trials being conducted in the developing world. This tension over the need for a protocol, and more particularly how it should be operationalized, continued throughout the negotiations.

Central Issues in the Protocol's Negotiation
There were four major issues within the Protocol's negotiation: (1) the Protocol's scope, (2) trade issues, (3) criteria for decision making, and (4) exporter responsibilities. This section outlines debates on each issue as well as the final outcome.

Scope of the Protocol   Article 4 (Scope) defines the LMOs to which the Protocol applies. It specifically states, "This Protocol shall apply to the transboundary movement, transit, handling and use of all living modified organisms that may have adverse effects on the conservation and sustainable use of biological diversity, taking also into account risks to human health." A central question within the negotiations was how the Protocol would address LMOs that are internationally traded commodities. While most delegates recognized the need to focus on LMOs that would be directly introduced into the environment, many countries, such as those in the Miami Group, did not want a system that would require importing Parties to go through an extended decision-making procedure for each LMO to be shipped, particularly those intended for other purposes (e.g., LMO-FFPs, LMOs in transit and for contained use, and pharmaceuticals). Others countered that states should have the right to make decisions on all LMOs, given differences in natural environments as well as the reality that LMOs designed for food could also be used for planting. Despite resistance throughout the negotiations, the Miami Group eventually agreed that LMO-FFPs could be included within the Protocol's scope, while calling for their exemption from the standard decision-making procedure for direct introductions. This led to the development of an alternative procedure for LMO-FFPs, which reduces timeframes and removes the burden on the exporting Party to ensure receipt of an approval to export, while still allowing importing Parties to assess risks and approve imports upon their own initiative.

As referenced above, three additional categories were debated: LMOs for transit, contained use, and human pharmaceuticals. LMOs in transit are transported through a state's national jurisdiction (by land, air, or sea) with their final destination being another state. LMOs destined for contained use will be transported to a facility or installation, which effectively limits their contact with the environment. Finally, another area of debate was whether to include pharmaceuticals for humans. Given that all three of these categories of LMOs are not intended for direct introduction into the environment, numerous countries from the Miami Group, European Union, Compromise Group, and the CEE argued that they did not pose a threat to biodiversity and therefore should not be subject to burdensome approval processes. Representatives of the Like-Minded Group countered that despite best intentions, accidents and unintentional releases do occur, so states should have the sovereign right to make decisions regarding what crosses their borders. The final decision in each case was to exempt these categories of LMOs from the advance informed agreement procedure while recognizing the right of Parties to regulate the transport of LMOs through their territories, set standards for contained use within their national jurisdictions, and subject all LMOs to risk assessments prior to making decisions on import.

Trade Issues The negotiations raised two primary issues regarding aspects of the international trade regime: the Protocol's relationship with other international agreements and non-Parties. These issue areas were manifested during specific legal discussions on how the Protocol and its provisions would relate to existing trade rules and obligations.

During the negotiations delegates debated whether specific language should be included to clarify the Protocol's relation to other international agreements, such as those negotiated under the World Trade Organization (WTO), in cases of conflict.[3] To clarify the potential ambiguities, the Miami Group supported including a separate article stating that the Protocol's provisions would not affect the rights and obligations of Parties under any existing international agreement. Such a provision is not uncommon in international law, and is generally referred to as a "savings clause" as it "saves" preexisting obligations. This would confirm that WTO rules trump the Protocol in the event of a dispute between countries, particularly on issues related to unwarranted discrimination between LMOs and their non-GM counterparts. Some countries suggested modifying this proposal by including an exception for cases of serious threat or damage to biodiversity. The Like-Minded Group, later joined by the European Union, strongly opposed inclusion of a savings clause in the operative text of the Protocol, and instead favored addressing the issue in the preamble. While technically no less legal, preambular provisions are generally viewed as statements of intent and are not given the same weight as obligations contained within the operative text of an agreement.

Given the two almost intractable positions, which contributed to the collapse of the Cartagena meeting, work at the Vienna informals and the final Montreal meeting sought to craft text reflecting the general principles of both positions. This resulted in three preambular statements related to trade, the Protocol and other agreements:

• Recognize that "trade and environment agreements should be mutually supportive with a view to achieving sustainable development."

• Emphasize that "this Protocol shall not be interpreted as implying a change in the rights and obligations of a Party under any existing international agreements" (this reflects the pro-savings clause position).

• Understand that "the above recital is not intended to subordinate this Protocol to other international agreements" (this reflects the position of those that did not want the protocol subordinated to the WTO).

Regarding the Protocol's application to non-Parties (those countries not ratifying the agreement), two main concerns arose during the negotiations. First, how will Parties to the Protocol trade with non-Parties, especially in the initial years after the Protocol comes into force when there are still a small number of Parties? The second concern specifically relates to the United States, which is the world's leading exporter of LMOs and has the most developed biotechnology industry. Hence the US position with regard to trade, as well as LMO exchanges for research and other purposes, could not be disregarded despite the fact that the US is not a Party to the CBD (which is a requisite for becoming a Party to the Protocol). Discussions also touched on the issue of nondiscrimination and how to ensure that non-Parties do not enjoy any advantages over Parties to the Protocol in the trade of LMOs.

At BSWG-3 and BSWG-4 many developing countries supported a ban on trade with non-Parties, and there was significant debate over whether this ban would serve as an incentive or disincentive for non-Parties to join. The call for a ban was gradually dropped by BSWG-6 as the existing economic realities and pressure from developed countries took hold. Then the language in the Protocol moved toward ensuring that the exchange of LMOs with non-Parties would be consistent with the Protocol's objective, which is reflected in the final text of Article 24 (Non-Parties).

Decision-making Criteria   States debated criteria and information that should be applied and assessed when making decisions regarding requests to import LMOs. The basis of the Protocol's advance informed agreement procedure allows for scientific risk assessment of potential harm. However, some groups wanted to expand this to include precautionary action where scientific evidence of harm is uncertain or where there may be negative socioeconomic consequences. Such concerns were seen to extend beyond the bounds of what was currently allowed under the WTO (specifically its Agreement on the Application of Sanitary and Phytosanitary Standards—the SPS Agreement), which led to debates on whether the additional criteria could constitute disguised barriers to trade.

Thus one of the most controversial issues during the negotiations was whether and how the precautionary principle should be included within the Protocol's text. The principle provides general guidance on how to act under conditions of scientific uncertainty regarding potential threats to the environment or human health (see Freestone and Hey 1995; Sands 1994, 208–13). The principle is a relatively new concept in international law with its first notable inclusion in Principle 15 of the Rio Declaration on Environment and Development, which states: "Where there are threats of serious or irreversible damage, lack of full scientific certainty shall not be used as a reason for postponing cost-effective measures to prevent environmental degradation" (United Nations 1992d).

The principle's application arguably presents the biggest challenge to trade rules under the Protocol, as it provides the legal basis for countries to ban imports for fear of adverse environmental or health implications. With potential profit margins for the biotechnology industry in the billions, the Miami Group sought to reduce such uncertainty at the policy level by placing reference to the principle within the Protocol's preamble and/or objective and ensuring that all decision making be based on sound science and rigorous risk assessments (Cosbey and Burgiel 2000). In their statements the United States and Canada argued that the Protocol itself is an expression of the principle. The European Union and Like-Minded Group preferred more specific formulations within the Protocol's operative section that would allow greater flexibility in using the principle. The final text refers to the precautionary approach and Principle 15 of the Rio Declaration in the preamble and in Article 1 (Objective). More important, the Protocol includes language allowing precautionary action within provisions on decision making for LMO-FFPs and LMOs intended for introduction into the environment (contained in Articles 10.6 and 11.8).

Another heavily debated criterion for making decisions on imports under the Protocol was the consideration of socioeconomic factors, such as would allow Parties to base their decisions to ban imports on nonscientific factors. Part of the debate centered over the ambiguity of the term and its potentially vast coverage, including environmental impacts on farmers' livelihoods, economic impacts from import substitutions, corporate monopolization and vertical integration, as well as moral and religious beliefs. Generally, developed countries, especially the Miami Group, did not want to include such a provision as it could prove even more onerous than the precautionary principle in establishing trade barriers against LMOs. The Like-Minded Group supported such a criterion noting a vast array of potential impacts on farming communities, traditional agricultural practices, national food security, and increased competition with non-GM export crops.

By the Vienna informals socioeconomic factors had been removed from the list of core areas under consideration, and they were relegated to informal consultations, thereby limiting discussion during the final stages of negotiation. At the conclusion of the negotiations, Article 26 (Socioeconomic Considerations) remained within the Protocol almost as a *fait accompli*. The article states that Parties, when making decisions regarding imports, can take into account socioeconomic considerations arising from the impact of LMOs on the conservation and sustainable use of biodiversity, although such considerations should be consistent with Parties' other international obligations.

Exporter Responsibilities   The final area of conflict within the negotiations concerned provisions on liability and redress and on documentation, which presented additional actual and potential costs for exporters. A liability and redress provision would hold exporters responsible for illegal acts or adverse impacts of their activities on the environment and human health. Over the course of the negotiations, numerous developed countries and industry representatives objected to the inclusion of such a provision stating that national liability and tort laws should be applied, while also highlighting the complexities and long-term timeframe needed to develop an operational mechanism for determining liability. They further noted that assessing monetary damages to the environment, human health, and restoration costs would be extremely difficult. The Like-Minded Group and ENGOs supported inclusion of a liability and compensation provision, arguing that without such "teeth" promoters of biotechnology would be less inclined to abide by the Protocol and could not be held fully accountable for their actions, especially in countries with weak or no liability legislation.

Although raised from the start of negotiations at BSWG-1, the specifics of such a system were never substantively debated. Without agreement by the final BSWG meetings, delegates acknowledged that there would be insufficient time to detail such a mechanism and ultimately agreed to consider the issue after the Protocol's entry into force. Article 27 (Liability and Redress) states that the first meeting of the Parties shall develop a process for the "appropriate elaboration of international rules and procedures in the field of liability and redress for damage resulting from transboundary movements of LMOs" to be completed within a four-year timeframe.

Requirements for documenting shipments of LMOs were the final issue to be resolved at the resumed ExCOP in Montreal. Documentation refers to the information accompanying a shipment and was distinguished early on from the actual labeling of packaging for consumer products. The most significant obstacle to agreeing on how to document shipments of LMOs, most specifically LMO-FFPs, was whether such action would ultimately require segregation of GM and non-GM goods. The Miami Group and industry argued that segregating GM crops would require duplicating the existing collection and distribution systems, which could cost billions of dollars and would not be feasible in the near term (Bullock, Desquilbet, and Nitsi 2000). Others argued that a Protocol designed to provide advance informed agreement on LMO shipments and to improve information sharing on LMO-FFPs should at minimum specify the type and amount of LMOs in a shipment. The final compromise on documentation for LMO-FFPs in Article 18 (Handling, Transport, Packaging, and Identification) calls for identifying such shipments as "may contain" LMOs and specifying that they are not intended for introduction into the environment. The ExCOP also agreed that the meeting of the Parties would decide on further requirements, including specification of identity.

Non-state Actors

Whereas a number of studies have focused on the impact of ENGOs in multilateral environmental negotiations or the role of industry groups in trade negotiations, few have looked at the role of these traditionally opposed groupings within the same negotiation process. The number of ENGOs and industry representatives participating in the negotiations increased with each meeting despite the restrictions that were placed on their involvement over time. Industry participation was initially far

Table 4.1
ENGO and industry groups at negotiating sessions

Meeting	ENGOs	Industry
BSWG-1	20	10
BSWG-2	17	6
BSWG-3	14	14
BSWG-4	20	22
BSWG-5	17	31
BSWG-6/ExCOP	39	23
Vienna informals	16	23
Resumed ExCOP	72	42

Note: The numbers correspond to the number of registered organizations and not to individual participants, nor do they include registered universities or media. For more details on participant lists, see Burgiel (2002).

below that of ENGOs, but over the course of the negotiations increased and, at times, exceeded the number of ENGO groups (table 4.1). For the Protocol, ENGOs and industry held opposing views on most of the major issues under debate. A simple zero-sum assumption would be that one side's success in influencing the negotiations would necessarily be a measure of the other side's failure. While such give-and-take oversimplifics the case, it is a useful dynamic to consider, especially when asking whether such efforts might offset each other, or when examining how ENGO and industry interests aligned with the positions of different negotiating blocs of states.

This section outlines the positions and tactics ENGOs and industry in the Protocol negotiations. To maintain continuity, each group will be analyzed separately, starting with an overview of their participation in the negotiations including specific organizations involved and their general orientation, including priority agenda items and framing of the major issues. The analysis then focuses on ENGO and industry positions on the central issues. Finally, the section focuses on the ability of ENGOs and industry to influence the negotiations and the factors conditioning their success or failure. In evaluating influence Corell and Betsill (chapter 2) suggest the use of process tracing to specifically link ENGO positions and actions directly to changes within the behaviors of countries/

negotiating blocs and the course of negotiations. Within the discussions below there are no obvious examples of key language or concepts injected into the Protocol's text that can be used to trace NGO influence through to the final outcomes. The use of counterfactuals can best be used to analyze how non-state actors and negotiating blocs intersected.

ENGOs

Participation   ENGOs represented a diversity of groups, ranging widely in size and geographic distribution. Larger ENGOs and ENGO networks such as the Third World Network (TWN), Greenpeace, Worldwide Fund for Nature International (WWF), and Friends of the Earth International (FOE) worked together with smaller ENGOs and ENGIs (environmental nongovernmental individuals), such as Accion Ecologica, the Australian GeneEthics Network, Diverse Women for Diversity, Ecoropa, the Edmonds Institute, the Council for Responsible Genetics, and the UK Food Group. ENGO support for a protocol extended back to the initial discussions under the CBD.

During the early course of negotiations, the ability of both ENGOs and industry to access the formal discussions both in written and verbal form was generally open with a few exceptions (e.g., a closed contact group session at COP-2 to decide on whether to negotiate a protocol). However, as with other negotiating processes, the number of opportunities to access the floor was limited for non-state observers, and therefore ENGOs generally worked collaboratively to maximize input through joint statements, other lobbying activities, and position papers.

In 1998 at BSWG-4, the BSWG Bureau took a decision stating that ENGOs and industry groups could only participate as observers without the right to speak or intervene in the negotiations. Previously such groups were able to make statements during most official sessions. However, from BSWG-4 onward, these groups were limited to statements during opening and closing plenaries. At BSWG-5 (Montreal 1998), further restrictions were placed on ENGOs and industry representatives, barring them from initiating direct contact with delegates (either orally or through written materials) during formal negotiating sessions and allowing Co-chairs of working groups to restrict their access. ENGOs and industry could still interact with delegates but only outside of official sessions.

Additionally, from the close of BSWG-6 through to the final resumed ExCOP, discussions were conducted either as closed meetings of "friends of the Chair" or in the "Vienna Setting" with two representatives for each of the five major negotiating blocs. These formats also effectively prevented any direct ENGO and industry participation in formal sessions. Overall, such decisions and changing formats of the discussions limited the ability of non-state actors to participate and lobby actively in the later stages of the negotiations. At this time the remaining key issues were also highly defined and contentious, which further limited the ability of ENGOs to make major impacts.

In framing biosafety issues to the negotiators, media and public, ENGOs addressed a wider range of subjects than contained in the narrow mandate of the Protocol. Instead of looking solely at aspects of international LMO transfers, ENGOs highlighted a range of other concerns:

• Broader technological and social issues (e.g., genetic use restriction technologies and food security)

• Information on relevant national events and legislation (e.g., illegal or undocumented imports and introductions, national bans on LMOs, and labeling)

• Specific varieties of GMOs (e.g., *Bt* maize and genetic use restriction technologies)

• Impacts on human and environmental health

• Other policy issues (e.g., industry and government collusion and impacts on public sector science)

As observers, ENGOs had greater freedom to comment on a variety of issues, often using simpler, more direct language than the legalese of government officials. This broader approach provided greater recognition and resonance to observers outside the negotiations, while also simplifying the details of complex legal and regulatory negotiations. This expansive frame gave ENGOs a position from which to request inclusion of elements like product labeling and socioeconomic considerations that otherwise seemed beyond the Protocol's scope.

This broad range of interests also related to ENGOs' approach to agenda setting. While their ability to influence the actual negotiating agenda was limited, through their statements and position papers ENGOs took a proactive stand on what should be included on the agenda (and thereby the Protocol), particularly with regard to liability, the precautionary principle, and a comprehensive scope including all LMOs. This "positive listing" approach contrasted with industry's "negative listing" approach, which sought to keep particular issues off the agenda and out of the negotiations. As will be discussed below, such a positive approach was more viable during the earlier, and less politicized, rounds of negotiations.

Positions on Central Issues   Over the course of the negotiations, ENGOs were the most supportive of a strong and comprehensive protocol incorporating provisions on liability, socioeconomic conditions, and the precautionary principle. They were also at the forefront arguing that the Protocol should not be subordinated to the WTO on biosafety matters. Positions on specific issues are detailed below.

Scope of the Protocol   From the start of the negotiations ENGOs generally supported an all-inclusive scope. They argued that given gaps in existing science, risks should not be taken by omitting particular subsets of LMOs from various provisions of the Protocol, especially as an LMO might have different uses and effects across countries and environments. They further noted that an inclusive scope provides more control at the national level regarding what can be imported. Only a few groups had more specific comments regarding the inclusion of LMO-FFPs. At BSWG-4, Greenpeace International argued that excluding LMO-FFPs would exempt the vast majority of new biotechnology products, which are generally traded in processed form, thereby eliminating a crucial subset of LMOs from the Protocol's provisions. TWN argued that processing does not render LMOs harmless, since recombinant DNA, toxic residues, and antibiotic resistance marker genes can still exist in quantities sufficient to pose a threat to the environment or human health. FOE objected to the proposed exclusion of LMO-FFPs and to the development of an alternative procedure for their import, stating that it would not always be possible to distinguish between LMOs imported for introduction into the environment and those imported as commodities. Specific comments on limiting obligations on LMOs for transit, for contained use and as pharmaceuticals for humans were limited in view of general ENGO support for an inclusive scope.

Trade Relations   ENGOs presenting position papers or making statements unanimously supported the Protocol's precedence over the WTO and international trade rules. Some saw the issue of biosafety as distinct from trade and thereby stated that any trade-related issues be dealt with in other forums, whereas others, noting an implicit linkage, stressed the need to ensure that the Protocol was preeminent on biosafety issues. The level of detail within ENGO positions varied with the majority providing simple one-line statements of their position to a few groups providing a more detailed analysis of trade-related issues. Some ENGOs, such as the Royal Society for the Protection of Birds, argued that according to the rules of the Vienna Convention the Protocol should be superior to WTO rules as the Protocol would be more specific to the field of biosafety. Greenpeace International provided the most detailed analysis of a savings clause, noting that its inclusion would compromise the right of states to make sovereign decisions based on assessments of threat to their biodiversity.

ENGOs made only selective comments on the issue of non-Parties. The position of those providing comments early in the negotiations was for banning all trade with non-Parties to provide them with an incentive to ratify the Protocol. Some also argued that non-Parties should not be accorded rights and privileges to trade in LMOs without assuming the Protocol's responsibilities. Greenpeace International was the only group to call for a ban throughout the negotiations while others took a modified approach, stating that non-Parties should not be given a comparative advantage in the trade of LMOs.

Decision-making Criteria   ENGOs supported inclusion of the precautionary principle throughout the negotiations, often arguing that it should serve as the basis for a protocol concerned with safety issues. They linked the principle to provisions for risk assessment, risk management, socioeconomic considerations, and an inclusive scope. Additionally they challenged the Miami Group's distinction between sound science and the more subjective precautionary principle, stating instead that the two are mutually supportive. ENGOs argued that political, health, and other nonscientific decisions are made within risk assessments and should be explicitly recognized and not disguised under the assumptions embodied in "pure science."

More specific statements by a range of groups including FOE, Greenpeace International, Royal Society for the Protection of Birds, TWN, and WWF highlighted the principle's wide use in numerous international agreements, including the CBD, the UN Framework Convention on Climate Change, Principle 15 of the Rio Declaration, the WTO's SPS Agreement, and the treaty establishing the European Community. They also noted its use in national legislation and decision making in countries such as Austria, Canada, Luxembourg, and Norway. In addition they pointed to the difficulties in predicting ecological and health impacts, particularly for environments other than those where tested and for centers of genetic diversity and origin. Some other considerations were the absence of baseline data and long-term monitoring information, and the need to allow for global bans or phase-outs of specific LMOs or LMO traits and characteristics.

ENGOs favored including socioeconomic factors within the risk assessment process under the Protocol. Concerns included impacts on human health, human rights, food security, poverty reduction, indigenous and local communities, traditional forms of agriculture, and crops replaced by or competing with their GM counterparts. WWF and the Council for Responsible Genetics expressed their concern that widespread application of GM crops, especially as promoted by large transnational corporations, could undermine national and local economic activities and self-sufficiency in food production. Consequently ENGOs worked closely with developing countries to outline potential impacts on agricultural trade in non-GM crops and on small farming communities.

Exporter Responsibilities   From the start of the negotiations, ENGOs supported liability and compensation provisions. Even when the majority of countries questioned or opposed their inclusion, ENGOs worked hard with sympathetic developing countries to keep liability and redress on the negotiating table (Grolin 1996). Such efforts were most noticeable in the proliferation of buttons, stickers, and campaign materials stating "No Liability, No Protocol." ENGOs argued that without language on liability, the Protocol's enforcement would be extremely weak and that such provisions should model the polluter-pays principle and be an incentive for exporters to ensure safety. They countered arguments against a liability provision by stating that if LMOs are inherently safe, as argued by some countries, then those countries should have no concern over including an article on liability and compensation. ENGOs generally supported civil, state, and strict liability, which would cover harm to biodiversity, human health, and socioeconomic impacts (see Nijar 1997). Finally, regarding efforts to postpone discussions on liability, some ENGOs argued that experience with delaying liability provisions under other multilateral environmental agreements, such as the Basel Convention, had compromised their overall effectiveness.

On documentation, ENGOs supported mandatory identification of LMOs under the Protocol combined with segregation of GM from non-GM products during handling, transport, and storage. Segregation was seen as necessary to prevent any cross-contamination because tracking of LMOs under a system without segregation would be extremely difficult. Clear documentation would allow for tracing the chain of custody and increasing transparency, while also guaranteeing the consumer's right to know the contents of a product. Regarding critiques that segregation would be financially prohibitive, some ENGOs pointed to national examples, such as Iceland, where suppliers have segregated GM and non-GM products. Greenpeace also put forward detailed comments arguing that demand for GM-free products would provide a market incentive for segregation.

Influence Within their strategizing, ENGOs sometimes refer to engaging in insider versus outsider tactics to achieve objectives and assert pressure. Insider strategies involve providing direct commentary on negotiating texts, distributing scientific information, and lobbying government delegates. However, insider strategies are often criticized as being vulnerable to co-option by the process, as ENGOs are limited to working within (and thereby accepting) the existing framework regardless of how dissatisfactory it otherwise might be. Outsider strategies generally involve bringing public pressure and media exposure to problematic elements or delegations within the negotiating process. Such tactics often generalize or oversimplify technical legal points for the sake of communicating a broader message. Over the course of the biosafety negotiations ENGOs employed both strategies.

Overall, ENGOs had moderate influence on the Cartegena Protocol negotiations according to the criteria introduced in chapter 2 (table 4.2). While ENGOs had limited ability to directly shape the negotiating agenda, using an insider approach, they were able to supply governments with proposed text. However, it is unlikely that the text was taken into serious consideration, particularly at the close of the discussions where the blocs of states negotiated lines word by word. Thus the ability of ENGOs to directly influence the process became more limited over the course of the negotiations as the most contentious issues came to the fore. The following discussion will examine the interaction between ENGOs and three of the country negotiating blocs: the Like-Minded Group, the European Union, and the Miami Group.

ENGO/Like-Minded Group   ENGO positions generally mirrored those of developing countries, specifically the Like-Minded Group of developing countries, which also argued for a strong protocol. In terms of direct involvement, ENGOs were arguably most effective at the early stages of the BSWG. At this time they maintained a proactive orientation working with developing countries to keep issues on the table, such as labeling, liability, and the precautionary principle. In retrospect, this was most important for developing countries, which were fractured and disorganized and the only countries supporting these concepts. Additionally many developing country representatives, often single-person delegations, were learning about the issues as the negotiations proceeded. ENGOs were able to help these delegates understand the broader context of issues outside of their expertise.

Given that ENGOs provided a significant amount of information to developing country delegates, the absence of ENGOs would suggest that developing countries would not have been as well informed. The information was particularly crucial in the early stages of the negotiations where there was no unified developing country position and many delegates were still getting up to speed on the issues. At this point the European Union was opposed to a number of key issues, which they later came around to support. Through persistent advocacy and provision of detailed policy analyses, one could argue that ENGOs played a role in keeping liability and socioeconomic considerations on the negotiating table. As previously mentioned, at the start of the negotiations most countries did not support inclusion of these items under the Protocol. One could reasonably argue that without persistent ENGO support of these issues during the initial meetings of the BSWG, they most likely would have been take off the agenda. Only in the mid to late stages of the negotiations did developing countries (which then evolved into the Like-Minded Group) take up the charge on these issues. Also by this time the European Union had become more amenable to addressing liability and socioeconomic issues.

ENGO/EU Relationship   To learn more generally about the food safety issues within Europe, the ENGO/EU relationship needs to be examined outside the context of the Protocol negotiations. Through an outsider approach ENGOs exerted significant effort in raising public concern about GM food safety issues in order to influence politicians within Europe. For example, a number of environmental ministers directly criticized Pascal Lamy, the EU Trade Commissioner, for the pro-biotechnology stand that he took at the Seattle WTO Ministerial, which was seen as contradicting the overall EU position. The European position changed notably during the course of negotiations to a much more precautionary approach. It is possible to trace the shift in EU positions in the BSWG with this broader shift in politics around food safety more broadly in Europe. For example, outbreaks of mad cow disease and dioxin-tainted food in the 1990s spurred the public, consumer groups, and ENGOs to be more critical of national and regional policies in the European Union. Within this context, ENGOs highlighted growing concern over LMOs and their unknown health and environmental effects, making genetically modified foods another high profile issue as was initially seen in national bans by countries such as Austria, France, and Luxembourg on particular GM varieties (Mann 1999). While establishing the argument that ENGOs themselves caused this shift (and making a further linkage with the biosafety negotiations) is beyond the scope of this piece, a number of scholars have documented the role and

Table 4.2
ENGO influence in the Biosafety Protocol negotiations

		Evidence		NGO influence? (yes/no)
	Influence indicator	Behavior of other actors…	…as caused by ENGO communication	
Influence on negotiating process	Position of key states • Like-minded group	• Positions on most issues aligned closely with ENGOs.	• ENGOs provided policy analysis and support country delegations.	Yes
	• European Union	• Position shifted over course of negotiations to more closely align with ENGOs.	• ENGOs publicized debate about the safety of GMOs in national contexts.	Yes
	Agenda setting	• Liability and socioeconomic considerations stayed on the agenda despite general opposition outside of developing countries in early stages of negotiations.	• ENGOs showed persistent advocacy.	Yes
Influence on negotiating outcomes	Final agreement/procedural issues	• Canada and US allowed for conclusion of negotiations.	• ENGOs engaged in shaming and holding countries accountable if talks collapsed.	Yes

Final agreement/substantive issues			
• Scope	LMO-FFPs were subject to an alternative process; contained use, transit, and pharmaceuticals are generally exempted.	• ENGOs wanted all included.	No
• Trade issues	Ambiguous preambular language was used with relation to WTO.	• ENGOs wanted Protocol to be superordinate to WTO.	No
	• Trade was allowed with non-Parties in accordance with Protocol's objectives.	• ENGOs wanted trade with non-Parties prohibited.	No
• Decision-making criteria	• A precautionary principle was included.	• ENGOs wanted precautionary principle to be included.	Yes
	• Socioeconomic considerations were included.	• ENGOs wanted such considerations to be included.	Yes
• Exporter responsibilities	• Liability and redress provisions were included.	• ENGOs wanted a detailed liability mechanism.	No
	• Limited documentation was required.	• ENGOs wanted all LMOs labeled.	No
Level of Influence			Moderate

influence of groups like Greenpeace, FOE, and WWF in European discussions over food safety and GMOs (Ansell, Maxwell, and Sicurelli 2003; Arts and Mack 2003: 30; Bail, Decaestecker, and Jorgensen 2002: 173; Bernauer and Meins 2003).

ENGO/Miami Group   ENGOs' primary approach to the Miami Group was to "shame" them into adopting a protocol. With largely divergent positions, ENGOs portrayed the Miami Group as counter to environmental and public interests and responsible for the collapse of the first ExCOP in Cartagena. The Resumed ExCOP took place only two months after the WTO's Seattle Ministerial, which had direct implications on the Protocol's negotiations. During the WTO's Seattle meeting, Canada, Japan, and the United States supported development of a working group under the WTO to address trade and biotechnology issues. ENGOs portrayed this as an attempt to preempt the conclusion of the negotiations on the biosafety protocol, and later attributed the collapse of the meeting to the unreasonable demands of the United States and other industrialized states. ENGOs used this as a drastic backdrop for challenging countries, particularly the United States and Canada, to contribute to the failure of yet another set of negotiations.

This tactic is also evident in the approach of ENGOs toward the Canadian government and minister. At the start of the Resumed ExCOP, David Anderson, Canada's Minister of the Environment, was not expected to attend. The public relations campaign that ENGOs, particularly those with Canadian constituencies, embarked upon sought to portray this as the host country's disdain for the Protocol and biosafety issues, particularly because a large number of other environmental ministers were expected. Minister Anderson's decision to attend the close of negotiations can largely be attributed to this campaign (Gale 2002: 259–60; Tapper 2002: 270; Arts and Mack 2003: 27).

Such public pressure was not necessarily directed at any single issue, but it was arguably significant enough to keep the United States and Canada at the negotiating table. If ENGOs had not been present, particularly in the final stages of negotiations, Canada and the United States would not have had as much pressure to allow for the successful conclusion of the Protocol. Instead, they potentially could have prolonged negotiations, which accorded with the stated American view that "no protocol is better than a bad protocol."[4]

Industry

Participation   Although industry participation in CBD meetings is generally low, their representation in the biosafety talks increased dramatically over time (Grolin 1996). Representatives generally came from industry associations, such as the Biotechnology Industry Organization (US), BIOTECanada (Canada), EUROPABIO (Europe), the Global Industry Coalition, the Green Industry Biotechnology Platform (GIBiP), Japan Bioindustry Association, American Seed Trade Association, the International Chamber of Commerce, Grocery Manufacturers of America, the US Grains Council, as well as from some of the larger biotechnology and seed companies, such as Cargill, DuPont, Merck, Monsanto, Novartis, and Pioneer Hybrid. This increasing industry presence was a clear indication of the trade and economic interests at stake in the Protocol's negotiation.

On participation, industry representatives were subject to the same restrictions as ENGOs. However, generally, industry representatives were not as vocal or prolific as their ENGO counterparts in terms of distributing materials or making interventions during the earlier stages of the negotiations. Industry was also less likely to provide proposed text for negotiation or to attempt to influence the agenda, other than to oppose discussion of issues of direct concern to them (e.g., liability and labeling). Another fundamental part of their effort was countering the claims of ENGOs and others regarding the negative aspects and impacts of LMOs on the environment, human health, and socioeconomic well-being.

Industry representatives tended to serve as resources for delegates, explaining biotechnology issues and concerns and hosting lunchtime briefings and side meetings on the application of specific biotechnologies. These events often focused on the role of biotechnology in the developing world, and industry was also a strong supporter of other capacity-building efforts for developing country delegates. In effect industry maintained a lower profile than ENGOs. This also was because in countries with biotechnology industries, business representatives often had better access to government officials for inter-sessional dialogue.

In contrast to the ENGOs, industry representatives sought to narrow the mandate of the negotiations and to focus on transparent, science-based risk assessment and decision-making procedures, as well as information sharing and capacity building for developing countries. Some industry representatives also emphasized the need for a harmonized regulatory framework that would help streamline development of national legislative frameworks. They saw the Protocol as an instrument to enhance development of biotechnology with special attention to developing countries, while facilitating smooth and predictable transfer of LMOs among countries. Industry representatives wanted to avoid a protocol containing burdensome restrictions on biotechnology research and development, commercialization, product segregation, labeling, and more generally on trade in GM goods. Many industry representatives were also resentful of their portrayal as opponents of the Protocol, noting that it could be advantageous for commerce to harmonize disparate national regulations (deGreef 2000).

Thus industry emphasized the broader social and economic benefits of biotechnology. They argued for their contribution to sustainable agriculture and food production, in increasing crop yields, nutritional content, and food quality while reducing inputs and enhancing pest resistance and weed control. Benefits in the area of medicinal research were promoted as pharmaceutical production, cancer research, and animal vaccines. Finally, industry emphasized their contributions to capacity-building, emphasizing the role of biotechnology in industry development, training, bilateral research cooperation, national policy development, regional policy coordination, risk assessment and management, and public awareness.

By the end of BSWG-5 in 1998, industry representatives had organized themselves under the umbrella of the Global Industry Coalition (GIC), which presented consensus industry positions. The GIC includes approximately 2,200 companies in more that 130 countries, from sectors including agriculture, food production, human and animal health, and the environment. The GIC sought to limit the purpose of the Protocol to protecting biodiversity and to avoid negative impacts on trade. The GIC also stated that a poorly conceived protocol would undermine economic development by denying biotechnology's benefits to developing countries, compromise established scientific processes for evaluating LMOs, impede technology transfer and research cooperation, and hinder generation and sharing of biotechnology's benefits.

Thus in framing their priorities for the Protocol, industry worked to limit the number of issues and the scope of the negotiations to a strict interpretation of the original mandate contained in CBD Article 19.3. While ENGOs sought to include additional items, industry endeavored to keep issues such as liability, labeling, and the precautionary principle out of the debate and the Protocol as they would entail short- and long-term commercial costs. As the negotiations progressed and the provisions under discussion became more refined, industry delegates were better able to detail how proposals could impact international trade in GMOs. For example, in the later stages of negotiation American industry groups highlighted the difficulties in segregating GM and non-GM grains for documentation purposes, while also commenting on the practical difficulties of proposed timeframes and procedures for decision making regarding the various categories of GMOs. Such arguments were generally stronger toward the end of the negotiations, as the ramifications of the negotiating text became clearer.

Positions on Central Issues

Scope of the Protocol   From the start of the BSWG negotiations, industry groups consistently stressed that the Protocol should only apply to those LMOs with potentially adverse effects on biodiversity. The GIC argued that the Protocol should exclude LMOs not likely to present risks:

- Non-viable products of LMOs, such as processed foods and feeds
- Health care products and pharmaceuticals
- Products destined for contained use, such as for manufacturing and research
- LMOs in transit through a country
- Commodities not intended for deliberate release into the environment, such as foods for processing

Industry groups argued that inclusion of such nonhazardous products and activities would overburden countries' regulatory systems, thereby impeding commercial activity without compensatory environmental benefits. Alternatively, the GIC proposed that the Protocol and its decision-making procedure should apply to a small subset of LMOs that cross international boundaries and present potential adverse impacts to biodiversity. Thus industry stated as far back as BSWG-2 that LMO-FFPs should not be included within the Protocol's scope as such products are highly processed and thereby cannot reproduce or present an environmental threat. Representatives of the International Association of Plant Breeders (ASSINSEL) and GIBiP also argued that an exporter might have no knowledge of the exact composition of a shipment, given cross-pollination and/or mixing from the time of planting through harvesting, shipping, and processing, which would complicate strict adherence to the advance informed agreement procedure.

GIC also argued that LMOs in transit did not present a direct threat to the environment as there is no intention for release. For LMOs in contained facilities, industry representatives noted that hundreds of thousands of such transfers occur annually for purposes of commercial and academic research. Finally, acknowledging that pharmaceuticals are generally kept contained, industry representatives argued against including pharmaceuticals for humans to ensure access to medicines. Specifically, research on diseases particular to developing countries, such as malaria, hepatitis, and tuberculosis, could be impacted by import regulations, as well as those pharmaceuticals with limited shelf lives.

Trade Relations   Industry representatives generally favored the predominance of WTO rules over the Protocol to ensure the maintenance of fair and nondiscriminatory trade practices. During the negotiations industry representatives from the Biotechnology Industry Organization (BIO) and GIC stressed the need for rules to be compatible with existing international law, in particular, the GATT, and to be transparent and informed by scientific principles allowing for trade to proceed in a predictable manner. They also wished to avoid barriers to commercialization or investment in research and development activities and restrictions on the development of new products. Generally, industry wanted to prevent the possibility of countries applying trade-restrictive measures cloaked in ambiguous terms for environmental protection. GIC opposed the compromise preambular text proposed at the close of negotiations, stating that the formulation would subordinate existing trade agreements to the Protocol. Overall, industry, along with the Miami Group, favored inclusion of a savings clause, recognizing Parties' rights and obligations under preexisting international agreements.

On non-Parties, industry supported allowing trade with non-Parties and objected to any measures restricting such trade or other access. According to GIC, restrictions on non-Parties would have significant impacts not only on the trade of goods from biotechnology industries, particularly in the United States but also on research and exchange of materials between affiliates and collaborators in other countries.

Decision-making Criteria   Industry's views on the precautionary principle were intricately tied to relations with other agreements. Business representatives preferred the WTO's SPS Agreement prescriptions for risk assessment because they are based on sound science and form a more transparent and internationally harmonized basis for evaluating products. They feared that the proposed re-formulations would introduce a degree of uncertainty around national implementation and that countries might abuse the principle to absolve themselves of requirements to provide scientific evidence. Industry representatives favored harmonized national systems based on internationally accepted rules for risk assessment and sound science. A preferred approach using risk assessment and management would evaluate risks as or after they arise, as opposed to restricting commercialization or development before such risks are confirmed. It also presumes that the burden of proof will be on the party alleging damage, as opposed to the exporter having to prove that a product is risk free.

Similarly industry viewed inclusion of socioeconomic considerations within the Protocol's decision-making mechanisms as a potential means to refuse imports. A first concern was simply defining the scope of socioeconomic considerations, and industry groups argued that they should be addressed at the national level to encompass the range of different legal, political, and cultural systems. As the negotiations progressed, particularly in relation to the precautionary principle, industry argued that the Protocol should be grounded in science, inferring that socioeconomic considerations were nonscientific, and potentially political, criteria. As the GIC stated at the final ExCOP, "Including socioeconomic considerations in the decision-making process undermines the results of scientifically credible risk assessments and would surely create additional artificial barriers to trade" (Global Industry Coalition 1999). As on other issues, industry's positions on the precautionary principle and socioeconomic considerations coincided with those of the Miami Group.

Exporter Responsibilities   Industry generally opposed including liability and redress provisions in the Protocol. During the negotiations some industry representatives noted that biotechnology is currently subject to greater scrutiny than any other field of technology in human history and that such legal issues are better suited to other mechanisms. ASSINSEL and GIBiP also argued that liability should be considered under CBD Article 14 (Impact Assessment and Minimizing Adverse Impacts) and that national liability regimes are adequate and constantly adapted to emerging developments. Such arguments were reinforced at BSWG-6 and the final ExCOP, where industry representatives from the GIC argued that a liability regime based solely on a method of production would be unprecedented in international law and that liability issues should address the characteristics of a final product rather than the technology used to produce it.

The issue of documentation and labeling was a central concern of industry representatives, given the expense of segregation, testing, and documentation throughout production and transport, as well as a potentially negative public perception of labeled goods. Specific documentation requirements would force industries to develop separate collection, transport, distribution, and processing systems from field to market for GM and non-GM products, which would entail considerable expense. GIC called for simple requirements and recommended that any documentation requirements not exceed the negotiation mandate, while also noting that other international instruments, such as the Codex Alimentarius Commission, address documentation and labeling issues.

Given these concerns industry worked closely with the Miami Group to keep a documentation system deliberately broad (i.e., that shipments "may contain" LMOs). At the final negotiating session with the precautionary principle and trade issues resolved, many were surprised that the Miami Group raised this as an issue where they could not make further concessions. Detailed documentation requirements would entail actual, immediate costs to restructure all phases of the GM supply chain in the United States and other Miami Group countries, whereas application of provisions on the precautionary principle, socioeconomic considerations, or liability related to potential costs that may or may not be incurred in the future.

Influence   Industry groups had a moderate level of influence on the Cartegena Protocol negotiations; the observable effects of their participation are most apparent in looking at the negotiation process (table 4.3). Industry's underlying concern was to avoid undue restrictions or disincentives to the development and trade of LMOs and their products. Such concerns manifested themselves at different levels of the commercial development process and were linked to the Protocol's more controversial provisions. Generally, industry's position most closely aligned with that of the Miami Group. Given that the Miami Group included the only major GM producers in the world, it follows that the expression of such interests would accord significantly with the industries producing those goods. While industry interacted with all of the negotiating groups, including the European Union and the Like-Minded Group, their principal ally and point of leverage in the negotiations was the Miami Group.

Industry/Miami Group   As stated, industry operated from a more defensive orientation to keep onerous obligations off of the negotiation agenda. Perhaps most illustrative of this commonality of positions was the Miami Group's emphatic resistance to a strict identification and documentation system for LMO-FFPs, which was the final issue to be resolved under the negotiations. Miami Group representatives made the same statements and used the same rationale as their industry counterparts, who argued that a strict identification and documentation system would entail a multi-billion dollar restructuring of domestic systems (especially in the United States) for gathering, processing, transporting, and marketing GM and non-GM agricultural products. Another key area of

Table 4.3
Industry influence in the Biosafety Protocol negotiations

	Influence indicator	Evidence		NGO influence? (yes/no)
		Behavior of other actors…	…as caused by Industry communication	
Influence on negotiating process	Position of key states • Miami group	• Positions closely aligned with industry	• "Revolving door;" intersessional relations	Yes
Influence on negotiating outcomes	Final agreement/ substantive issues			
	• Scope	• LMO-FFPs subject to alternative process; contained use, transit, and pharmaceuticals generally exempted	• Industry decision-making procedure to apply only to LMOs for direct introduction into the environment	Yes
	• Trade issues	• Ambiguous preambular language re: relation with WTO	• Industry favored the predominance of the WTO	No
		• Trade allowed with non-Parties in accordance with Protocol's objectives	• Industry supported allowing trade with non-Parties	Yes

• Decision-making criteria	• Precautionary principle included	• Industry preferred risk assessment based on sound science	No
	• Socioeconomic considerations included	• Industry opposed inclusion of socioeconomic considerations	No
• Exporter responsibilities	• Liability and redress provisions included	• Industry opposed inclusion of liability and redress	No
	• Limited documentation	• Industry favored limited documentation	Yes
Level of influence			Moderate

concern was limiting the scope of the Protocol, particularly the application of its advance informed agreement procedure because it could significantly delay and restrict trade in LMOs. Industry positions again were matched with the Miami Group in an attempt to limit the procedure's application to only those LMOs for direct introduction into the environment, thereby waving onerous restrictions on LMO-FFPs, LMOs in transit and for contained use, and pharmaceuticals.

Another factor shaping industry influence was the supposed "revolving door" between the US government and major industry representatives, which is a potent form of insider politics involving the actual exchange of personnel. ENGOs were quick to document the high-level exchanges between industry and the US government and to associate this relationship with the almost identical slate of positions taken by the two groups. High-ranking US government personnel moving into the biotechnology industry included former Presidential advisors, Secretaries of Commerce and Agriculture, and administrators for the Environmental Protection Agency, the Food and Drug Administration, and the Animal and Plant Health Inspection Service. Corporations involved included BIO, Dow, DuPont, Monsanto, and Pioneer Hi-Bred (Burgiel 2002: 119).

Without the presence of industry at the negotiations themselves there would not have been as close scrutiny of how provisions under the Protocol would affect on the ground business practices. These repercussions were most clearly visible in the areas of the Protocol's scope, specifically around LMO-FFPs, and documentation and labeling issues. One could also argue that the positions of the Miami Group might have been less stringent, potentially resulting in a Protocol that was more protective of the environment and more restrictive to trade in LMOs.

Conclusion

From these discussions it is possible to conclude that ENGOs and industry did influence the negotiations on the Protocol to some degree. While levels of influence are relative, these findings suggest that non-state actors facilitated the negotiation process in both procedural and substantive terms. Their ability to influence the talks was contingent on a combination of tactics, timing, and objectives. In the early stages of the negotiations, ENGOs used a combination of insider politics working with developing country delegates to educate and to keep issues such as liability and socioeconomic considerations on the agenda. Alternatively, their insider push for more substantive discussion of liability or documentation and labeling later in the negotiations was severely limited. ENGO outsider tactics were arguably effective in building public pressure to help turn the EU position, and then such tactics played a role at the close of negotiations in pushing some Miami Group countries to stay at the table and complete the Protocol. In contrast, industry arguably exerted its most tangible insider influence in the middle and late stages by working with the Miami Group countries to limit the Protocol's scope and to keep detailed documentation and identification requirements off the agenda and out of the final text.

Analysis of how non-state actors influence negotiations requires in-depth and long-term observation, but researchers may not always have the opportunity to participate in each and every negotiating session. Refined methodologies and focal areas can ease this burden by pointing to the information and actions most relevant for study. Use of counterfactuals and process tracing, combined with analysis of information provision, opportunity structures, and key alliances, provides a range of tools for teasing apart complex negotiations to identify where ENGOs, industry and other non-state actors have been effective in the past and how they may increase their efficacy in the future.

Notes

1. LMOs are defined by the Protocol as any living organism that possesses a novel combination of genetic material obtained through the use of modern biotechnology (*Cartagena Protocol on Biosafety* 2000: 4).

2. For a detailed negotiating history, see coverage by the *Earth Negotiations Bulletin* at ⟨http://www.iisd.ca/biodiv.html⟩.

3. According to the Vienna Convention on the Law of Treaties, it is customary for the commitments of older agreements to supersede the obligations of a newer agreement, because the negotiation of a newer agreement should take into account the existing body and context of international law. However, if the newer

agreement addresses a specific issue and existing agreements only relate to that issue generally, then the provisions of the newer agreement should supercede those of previous agreements. The legal ambiguity rose around differing interpretations as to whether the Protocol's specificity or the WTO's preexistence would take precedence.

4. The point was made on a number of occasions by US negotiators, and is also reflected by Rafe Pomerance (2000: 6), a former US Deputy Assistant Secretary of State for the Environment.

5

NGO Influence in the Negotiations of the Desertification Convention

Elisabeth Corell

International environmental negotiations are often precipitated by increased environmental concern stemming from scientific findings that create reactions among decision-makers, the media, environmental organizations, and the public. An illustrative example is the discovery of ozone layer depletion, a phenomenon that attracted media attention and the subsequent international negotiations to curb the problem (Benedick 1991). The need to address environmental problems has enhanced the importance of those who can provide issue-specific information and advice to decision-makers. In the international context, diplomats and other government representatives work with numerous issues simultaneously, are sometimes poorly informed, and need expert advice to help them identify the policy options that coincide with their national priorities. This demand for specialized knowledge has led to an increased role for non-state actors in international environmental politics.

The role of non-state actors as providers of knowledge and expertise in international environmental negotiations has attracted significant academic attention. Specifically, the influence of two groups—formally appointed scientific advisers and nongovernmental organizations (NGOs)—on decision making at the international level has been the subject of several studies. Scientific expert groups are often appointed to advise diplomats in the preparation of negotiations, and the issues discussed at group meeting are often important keys to the origins of definition and the central understanding of the environmental problem being addressed (Boehmer-Christiansen 1994; Marton-Lefèvre 1994; UNEP 1998). Greater NGO access to international negotiations means that they too increasingly provide information and lobby for particular policy outcomes (Princen and Finger 1994; Potter 1996; Willetts 1996a; Clark, Friedman, and Hochstetler 1998).

This chapter examines the influence of environmental and social NGOs in the negotiation of the United Nations Convention to Combat Desertification (CCD), from the beginning of the negotiations in 1993 until the first Conference of the Parties (COP-1) in 1997. The data I draw on were obtained from official United Nations (UN) documentation, conference reports and newsletters, printed materials available at the negotiations, interviews, and my own observations. I collected these materials while attending all twelve negotiating sessions from 1993 to 1997.

The chapter begins with a background on how desertification has been addressed at the international level. In the following sections, evidence on NGO participation is presented, with attention to their activities, access to the negotiations, and resources. I use the analytical framework presented in chapter 2 to assess the level of NGO influence and discuss factors that shaped the ability of NGOs to influence the negotiations.

Desertification on the International Agenda

Severe droughts in the African Sudano–Sahelian region in the late 1960s and early 1970s prompted the United Nations to convene the Conference on Desertification in Nairobi, Kenya, in 1977. The main result of this conference was the nonbinding Plan of Action to Combat Desertification (PACD), which was to be implemented by the year 2000, with the United Nations Environment Programme (UNEP) having responsibility for its follow-up and coordination (United Nations 1980). However, implementation of the PACD largely failed and the issue of desertification re-emerged at the 1992 United Nations Conference on Environment and Development (UNCED). The initiative for a global convention on desertification originally came up at a 1991 meeting of African environment ministers in Abidjan, Côte d'Ivoire, in the preparation for UNCED. They felt that the interests of other regions of the world were being met by either the biodiversity or climate conventions (both of which were later signed at UNCED) and argued that the developing world, particularly Africa, needed something in exchange (Corell 1998). At UNCED, governments agreed to negotiate a binding legal agreement on desertification within one year.

Over the years the definition of desertification has been contested (Glantz and Orlovsky 1983; Blaikie and Brookfield 1987; Odingo 1990; Helldén 1991; Hare 1993; Mainguit 1994; Thomas and Middleton 1994; Middleton and Thomas 1997). For example, causal explanations range from human impact to natural (climatic) influences, or a combination of the two. A consensus definition of desertification was reached at UNCED in the negotiation of Chapter 12 of *Agenda 21*, which in turn appears in Article 1(a) of the CCD: "Desertification means land degradation in arid, semi-arid, dry sub-humid areas resulting from various factors, including climatic variations and human activities" (see United Nations 1992a). Desertification has social, political and economic facets and affects all regions of the world, although the problem is most severe in countries on the margins of the Sahara in Africa.

The Convention to Combat Desertification

The United Nations Convention to Combat Desertification (CCD) in those countries experiencing serious drought and/or desertification, particularly in Africa was negotiated during five intergovernmental negotiating sessions between May 1993 and June 1994 (INCD 1–5). The Convention entered into force in December 1996. Six interim negotiating sessions (INCD 6–10) were held prior to the first session of the Conference of the Parties (COP), which took place in Rome in September and October 1997.

The negotiations were deeply affected by North–South tensions. They addressed a number of issues, including commitments under the Convention; capacity-building; education and public awareness; national, subregional, and regional action programs; a special role for Africa; the "bottom-up approach;" a financial mechanism; the rules of procedure of the Convention; the relationship with other conventions; and the creation of a Committee on Science and Technology as the body for technical advice to the COP.

The CCD negotiations involved a number of actors: the Chairman, the Executive Secretary and the Secretariat staff, the negotiations Bureau, the International Panel of Experts on Desertification (IPED), and the members and observers of the Intergovernmental Negotiating Committee on Desertification (INCD). The INCD members, consisting of government delegations, made decisions regarding the contents of the Convention. Observers included intergovernmental organizations (IGOs); concerned specialized UN agencies, such as UNEP and the United Nations Development Programme (UNDP); and NGOs.

NGO Influence

This section applies the analytical framework developed in chapter 2 to assess the influence of NGOs on the CCD negotiations. I present empirical data on NGO participation to demonstrate how NGOs engaged in the negotiation process. I then examine evidence on NGO goal attainment to assess their effect on the negotiating process and outcome. Based on this analysis, I conclude that NGOs exerted a high level of influence on the CCD negotiations.

NGO Participation

NGOs were actively engaged in the CCD negotiations. From INCD-1 to COP-1 (1993–1997) a total of 187 environmental and social NGOs actively participated in the CCD process.[1] There was always a core group of about 40 organizations active at the meetings, and 30 organizations participated in five or more of the meetings. Of the 30, eleven were based in Africa, two in Asia, six in Europe, three in Latin America/ Caribbean, two in North America, and three in Oceania. In addition there were three "international" organizations with offices in several countries or that had representatives from different parts of the world attending different meetings. Africa was by far the most represented geographic region: almost one-half (91 NGOs) of the total participating NGOs and just over one-third of the organizations that attended most frequently were based in Africa. Among the NGOs attending the negotiations, most generally believed that African NGO interests should have priority.

The majority of the NGOs attending the CCD negotiations could be classified as representing grassroots interests, and many had little experience with international negotiations. As a result some of the early NGO coordination meetings were unfocused regarding strategy and priorities, and it took a while for the NGOs to develop a routine.[2] Fortunately, a number of more experienced NGOs devoted their time to supporting and providing know-how for the less experienced NGOs. The Environment Liaison Centre International, for example, prepared a "lobby manual" for NGOs involved in the INCD with the express purpose of providing "some practical tips on how to maximize influence in the negotiations."[3] As the negotiations progressed, the NGOs became increasingly experienced and their physical presence on the conference floor made it possible to follow the negotiations in detail, lobby delegates, and make relevant statements in the meeting.

At INCD meetings, NGOs coordinated their activities, usually met twice daily during the sessions, lobbied delegates, and held seminars. NGOs created their own working groups on issues pertaining to institutions; regional instruments for Africa and South America; capacity-building, education, and public awareness; financial resources and mechanisms; and science and technology (Walubengo 1994: 3–4).[4] This coordination enabled NGOs to make statements, which were often reproduced in *ECO*, a newsletter NGOs have published at environmental conferences since the UN Conference on the Human Environment in Stockholm in 1972, on behalf of all attending NGOs. NGOs at INCD meetings used *ECO* to analyze the negotiations from the NGO point of view, to inform delegates of the views and grassroots experiences of NGOs on desertification-related issues, and to inform fellow NGOs on progress in the negotiations.

NGOs generally tended to consult with government delegates from their own country or region and with those who spoke the same language, although they did interact with delegates from other regions, particularly during strong lobbying efforts on contentious issues or when they deemed it necessary to secure wide support for an issue they favored. As the negotiation process evolved, NGOs gained the trust and confidence of numerous government delegates and had the opportunity to meet formally and informally with delegations, allowing them to provide specific input on certain aspects of the negotiations.

NGOs were also active in between official negotiating sessions. They usually held coordination meetings before and after the INCD sessions, and in some instances, the NGOs held international or regional conferences in preparation for INCDs. Such conferences at times resulted in statements or reports, including drafting proposals, subsequently presented to the negotiations.[5] Also during the CCD negotiation process, NGOs created Le Réseau d'ONG sur la Désertification et la Sécheresse (RIOD)—a worldwide network for cooperation among NGOs involved in the implementation of the Convention. RIOD was established in June 1994, when the Convention text was agreed upon and subsequently gained recognition as an NGO focal point. The CCD Secretariat and governments took an immediate interest in this NGO network and the action program it adopted on its creation, demonstrating the respect and prominence these actors accorded to the NGOs involved with the INCD process. At the INCD meeting following the creation of RIOD, NGOs met daily with representatives from donor governments and UN agencies to discuss mechanisms for funding the NGO action program (Chasek et al. 1995). The Secretariat and governments have used RIOD to channel information to all NGOs interested in the Convention.

NGOs participating in the CCD process were given considerable access to the negotiations, due in large part to the precedent set at UNCED giving NGOs a greater role in international environmental decision-making processes. During the preparation of the UN resolution that established the INCD, the G-77 specifically stressed the importance of developing country NGO participation (*Earth Negotiations Bulletin* 1992). Paragraph 19 of that resolution subsequently invited "all relevant non-governmental organizations and especially ... non-governmental organizations from developing countries to contribute constructively to the success of the negotiating process" (United Nations General Assembly 1992). In this spirit NGOs were also strongly encouraged by the INCD Chair to participate in the negotiations. The Chair, as well as the Executive Secretary of the CCD Secretariat, took time at INCD meetings specifically to brief NGOs about the contentious issues in the negotiations, and to hear their views.[6]

According to one Secretariat staff member, the Chair and Executive Secretary "realized that things were moving in this direction as a result of the Rio process and there was already consensus that support should be given to the grassroots communities. They were the first to create a supportive atmosphere for the NGOs."[7] The Secretariat also took an active role in supporting NGO participation. They canvassed NGOs, particularly from developing countries, to attend the negotiations and organized workshops and contact group meetings between INCDs to promote NGO coordination. Most important, NGOs also received financial and other support. This, for instance, allowed them to hold meetings immediately preceding INCDs to prepare lobbying strategies.[8] To help NGOs from developing countries participate in the process, funding was provided through the NGO Unit of the CCD Secretariat for their travel and expenses at negotiating sessions (INCD document A/AC.241/ CRP.2).[9] Delegates in the INCD agreed that it was appropriate for the Secretariat to use some of its funds to sponsor NGO participation.

Nevertheless, some governments were initially quite critical of NGO attendance. Some delegates from African and Latin American governments privately objected to the close involvement of grassroots groups.[10] NGO representatives suspected that authoritarian governments in some African countries saw the NGO activities as a threat to their power (Simons 1994). To suppress this perceived threat, some countries included NGOs on their delegations in order to exercise control over potentially critical NGOs.[11] In one case, however, this did not prevent an NGO delegate from criticizing his industrialized country's policies, with the result that his government stopped sending him invitations to participate on the delegation. He then continued to attend the meeting as a regular NGO representative with observer status. On the other hand, some governments chose to include NGOs on their delegations in order to widen the base of decision making and provide a channel for the expertise and know-how of many NGOs. There was a good atmosphere for NGO lobbying; as one NGO representative noted, "lobbying is a two-way street, you need information from delegates that support your view about how to get your proposal accepted by other delegations."[12]

Technical knowledge was NGOs' most valuable resource in the CCD negotiations. In the eyes of desertification negotiators, NGOs possessed key know-how essential for effective treaty implementation and were referred to as "partners in development." The NGOs were perceived as the link between the international negotiations and affected local populations on the ground. In addition the NGOs represented what the

Convention refers to as "local/traditional knowledge," a type of knowledge recognized as an important complement to scientific knowledge for addressing dryland degradation (Corell 1999b). Throughout the negotiations NGOs conveyed local knowledge on a number of issues, including the realities of desertification for affected populations, successes and failures of development projects, and women's vital role in dryland management.

NGO Goal Attainment
The NGO position during the CCD negotiations can be summarized into three points: the agreement should (1) encourage the use of a participatory bottom-up approach in its implementation, (2) reflect the social and economic consequences of land degradation for populations in affected areas, and (3) provide "new and additional resources" for dryland management projects in affected developing countries. NGOs' participation in the CCD negotiations contributed to the development of an agreement that reflected this position (particularly points 1 and 2), and actions by the negotiators further indicate that the NGOs' position and expertise had been incorporated into and formed an important element of the negotiation process.

Effects on Negotiation Outcome   NGO influence on the CCD negotiations is reflected in the final text, which contains provisions consistent with many of the NGO goals. The CCD calls for a bottom-up approach and contains language on the social and economic consequences of dryland degradation. Throughout the negotiations NGOs supported the participatory bottom-up approach, which emphasizes the importance of participation by those affected by desertification in the development of antidesertification plans. This approach emanated from the lesson of failed development aid projects that without the participation of the local population, projects do not have any long-lasting effects.[13] Although new to international negotiations, using a participatory approach had been an objective in development aid for ten to fifteen years preceding the CCD process.[14] Supporting this approach was also in the interest of NGOs, since it could bring them further into the process of making and implementing policies. After all, compared to governments, NGOs have a significant ability to ensure that political commitment is turned into action at the local level.

The bottom-up approach to the CCD and the elaboration of National Action Programs (NAPs) was proposed by the African Group, receiving strong support from the INCD Chair, the Secretariat, IPED, and government delegations from all regions. However, the fact that it survived to become an important component in the final Convention text can partly be ascribed to NGO action. The centrality of the bottom-up approach for the implementation of the Convention might have been lost from the text without the constant pressure from NGOs. For instance, one NGO said that an important task for NGOs was to "keep repeating the phrase 'bottom-up approach' all the time to ensure that it does not get buried among all the brackets in the end."[15] According to the INCD Chair, "the NGO community made an essential contribution in constantly reminding the negotiators of the real issues at stake" (Kjellén 1994).

By encouraging and convincing delegates to keep the bottom-up approach on the negotiating agenda, NGOs were able to ensure the incorporation of numerous references to NGOs, popular participation, the importance of local/traditional knowledge and other NGO proposals in the Convention text. Notable is the language on the importance of local participation and NGO involvement in the implementation of the Convention, specifically in the NAPs, to combat desertification in Article 10, paragraph 2(f) (United Nations 1994). One government delegate said that this is "the only Convention where the NGOs are in the provisions to be involved in the implementation of the Convention. This is the first time in any international legal instrument."[16] The result is that in the Convention, NGOs are mentioned 29 times, including in the sensitive articles on financial mechanisms. By comparison, the text of the UN Framework Convention on Climate Change (UNFCCC), the CCD's "sister" convention that was adopted two years earlier in 1992, only mentions NGOs three times (United Nations 1992e).

Another example of NGO influence is the CCD recommendation in Article 21, paragraph 1(d) for the establishment of national desertification funds. The Environment Liaison Centre International presented this proposal on behalf of the NGOs at INCD-3 (Bernstein et al. 1994).[17] Although this language made it into the treaty text, it is debatable whether there have been any new resources devoted to dryland management activities after the Convention was adopted. NGOs therefore could be said to have influenced the theory of how to address the issue, but competing interests have continued to influence governments' decisions to allocate funding for the issue.

NGOs continued to press their concerns once the Convention was adopted. During discussions at INCD-8 about the CCD's subsidiary body, the Committee on Science and Technology, NGOs reportedly collaborated with governments and coordinated themselves in lobbying efforts to successfully add language on attention to women, local peoples, and traditional and local knowledge and technology to the decision (Corell, Mwangi, and Wise 1996). Each of these clauses further reinforced the NGOs' unique perspective in the language governing the conduct of the Convention's bodies.

Effects on Negotiation Process   In shaping the negotiating agenda on the bottom-up approach and on national desertification funds, the NGOs influenced the negotiating process in a number of ways that cannot readily be observed in the agreed text. For example, NGOs were able to open up new opportunities for participation as they gained the trust and respect of negotiators. At COP-1, an official Plenary meeting was dedicated to NGO dialogue for the first time in international negotiations (Corell et al. 1997b). The meeting was co-chaired by the Chair of the COP's Committee of the Whole and an NGO representative. This "breakthrough" inspired NGOs to propose that this arrangement continue, and the Argentinian delegation later presented a draft decision to include NGOs in the official program of future COPs (Corell et al. 1997a).

At the national level, previously suspicious governments became less wary of NGO activities as the process evolved, and they began to involve NGOs in antidesertification activities. The change of attitude was particularly noticeable in Latin America, where some NGOs had not even been able to meet with government representatives in their own country and had to perform their national lobbying at the international meetings. A few years later, however, the same NGOs were invited to government meetings that were specifically intended for communication and collaboration with the NGO community.[18]

Level of NGO Influence

Drawing on the analytical framework presented in chapter 2, my assessment is that NGOs exerted a high level of influence on the CCD negotiations. Throughout the process NGOs provided written and verbal information to the negotiation sessions and to some government ministries, and provided specific advice to and interacted with government delegations while present at the meetings. NGO delegates did not shape the initial framing of the desertification issue, but they decided to retain the language on the extent, impacts, and definition of desertification that had already been determined in the negotiations of chapter 12 of *Agenda 21* at UNCED—with virtually nonexistent NGO participation as very few were present. NGOs contributed to shaping the negotiating agenda by insisting on the necessity of a bottom-up approach to tackling dryland degradation and proposing the establishment of national desertification funds. Finally, they made important impact by ensuring that certain text was incorporated in the Convention.

This assessment is supported by the majority of the informants (negotiation participants and observers) who thought that NGOs, given their observer status in the negotiations, had had considerable influence over the outcome. One Secretariat staff member stated that the NGOs "definitely had influence."[19] Another staff member said the "negotiations were a breakthrough for the NGOs, who made sure they were part of every part of the process, and the INCD was open—as much as possible—for the NGOs."[20] Government delegates also viewed the NGO input positively, even as early as INCD-1. One delegate noted that the NGOs were "doing an important job" and another was "curious about what the NGOs will bring into the process. They seem to have gotten over the fuss that they usually have."[21] Without the NGOs, the Desertification Convention would not be the instrument it is today.

Conditioning Factors

NGOs were able to achieve a high level of influence on the CCD negotiations because of three factors: the link between the bottom-up approach and NGO participation in the implementation of the Convention, their homogeneous composition and interests, and the fact that NGO participation was encouraged by negotiators.

The Bottom-up Approach

The bottom-up approach created political space for NGOs to influence the CCD negotiations. The very first words in the Convention declare that human beings in areas affected or threatened by desertification are at its center (United Nations 1994a, preambular paragraph 1) and NGOs were welcomed in the negotiations as the representatives of these peoples. Negotiators continuously referred to the "grassroots," "the peoples living in marginal lands," and how the Convention should be designed to improve their living conditions. The NGOs were perceived as the link between the international negotiations and affected local populations on the ground. Indeed many NGOs in the CCD process considered themselves to be grassroots, or at least intermediaries between the INCD and the local communities, and "talked about national level projects as if they were the custodians."[22]

Substantial NGO participation was facilitated by the centrality of the bottom-up approach in the international negotiations—in themselves a truly top-down activity. Negotiators needed NGOs—regarded as representatives of civil society—to show that they were serious about the approach and to demonstrate that it was being used in the negotiations.[23] Moreover, whereas government delegates and other negotiation participants do have some practical experience regarding the environmental issue under negotiation, such knowledge often tends to be dated or of a general character. Thus there is a need for NGOs with up-to-date and in-depth practical experience to provide information and remind negotiators about the reality outside of the negotiating room. The level of importance attached to NGO participation in the CCD process is illustrated by the conclusion of a Secretariat staff member in that "at the end of the day, they will be implementing the Convention."[24]

NGO Composition

A second explanation for high NGO influence is the composition of the NGOs attending the CCD negotiations. The NGOs were a cohesive group with fairly homogeneous interests. While not all attending NGOs were grassroots organizations or from the African region, there was a sense that these interests should have priority. As discussed above, several NGOs with more experience in international negotiations devoted their time to supporting and providing know-how for the less experienced groups. This was an important element in the cohesiveness of the NGOs, since the former group very well could have pursued their own agendas rather than support their grassroots colleagues. This support was a sign of the extensive collaboration within the NGO community at the negotiations. As the grassroots groups became more experienced, all NGOs were better able to coordinate their efforts, eventually giving rise to RIOD.

Their relative homogeneity also allowed NGOs to present themselves to the negotiators as a single, coherent block, rather than a plethora of different interest groups. One Secretariat staff member stressed that this was smart of the NGOs to always present a united front. "Despite internal conflicts, whenever they met with the INCD Chairman it had to look good."[25] Most of the NGO interventions in the Plenary and Working Group sessions were made by one NGO representative speaking on behalf of all the NGOs attending the INCDs, often reading from a prepared statement. These unanimous statements were also circulated in the negotiating room or printed in *ECO*.[26] The NGOs facilitated this internal coordination and the production of statements during the meetings by setting up working groups on various issues that allowed them to issue position statements on short notice. These joint statements also facilitated communication with negotiators. Rather than needing to take multiple NGO views into account, there was only one statement for negotiators to relate to, which imparted a coherent and convincing impression. Single statements also made it easier for negotiators to integrate NGO views into the text under negotiation.

Additionally many of the international NGOs that first spring to mind as prominent actors in global environmental politics did not participate in the CCD process. One Secretariat staff member noted that there was "no Greenpeace, no WWF, no IUCN, no lawyers, no former diplomats, no experienced professional lobbyists."[27] Desertification was not a priority issue for most northern NGOs. However, this absence contributed to the cohesiveness of those NGOs attending the negotiations. The absence of large northern NGOs with their own agendas and political considerations made it easier for the participating NGOs to coordinate their activities. NGOs that had desertification as their central focus did not have to consolidate their views with representatives of large NGOs, which consider desertification to be only one issue on the wider environmental agenda. Moreover the participating NGOs had more of a common focus on the development aspect of the environmental problem of desertification and were not simply focused on remedying an environmental "harm." A Secretariat staff member noted that the majority of the NGO representatives attending the INCDs "are people with field experience or who live in desertified areas, and they have decided to address the problem thoroughly by coming to this international meeting."[28] Thus the desertification issue did not attract the large, and sometimes most aggressive, lobbyists, so the attending NGOs were able to build more of a relationship of confidence with the government delegations.

Last, no business NGOs were present to divert delegates' attention nor to provide alternative views, which are often in opposition to the environment and development NGOs. Industry participants often compete with environmental NGOs because industry seeks less stringent regulations and NGOs argue for more restrictive provisions. In the CCD case, however, there simply did not seem to be any business interests to protect in the worlds' drylands, therefore no need was seen to fund participation of industry representatives in the negotiations. Business interests were not put at risk by a convention that tackled a problem that was not perceived to be directly caused by industrial activities. One government delegate confirmed that the NGOs "were allowed to have more influence because industry wasn't there."[29]

Supportive Environment
The third explanation concerns the widespread support for NGO participation in the CCD process. In the words of one NGO representative, "Our success is not only a result of our hard work, it is also a result of cooperation with our allies in the Secretariat, some northern governments and a few southern governments."[30] As the *Earth Negotiations Bulletin*, a publication that followed the INCD negotiations from the start, noted in 1994 after the Treaty was adopted, "NGOs were extremely positive about the openness of the negotiating process and the extent to which they were able to influence decision making, especially around such issues as national desertification trust funds and NGO par-

ticipation in the development of national action programmes" (Bernstein et al. 1994).

Critics might suggest that the governments went along with the grass-roots emphasis because the negotiating states knew they were unlikely to live up to their commitments on paper. I argue that elements related to the timing of the talks, both as a follow-up to UNCED and a period when donors looked to NGOs as worthy aid recipients in the development arena, and a learning process about the role NGOs could play were responsible for the support NGOs received. These factors helped to further open up for political space for NGOs.

The negotiations began less than a year after UNCED was concluded in Rio de Janeiro, which had generated a positive attitude toward NGOs and their participation in international environmental decision making. In addition the prevailing trend among donors at the time of the desertification negotiations was to fund NGOs and NGO projects, rather than government-run projects. As donor countries assessed their records of successes and failures in development aid, factors such as corruption and top-down approaches to the implementation of aid projects led them to take a more critical stand on where to allocate new aid. Gradually they turned away from governments and toward nongovernmental and local organizations, whose resources can reach the intended people and areas through the bottom-up approach. The work of NGOs thus became a major feature of development policy. Donors poured funds into NGOs, governments allocated major responsibility to them, and NGOs increased in number and size (Hulme and Edwards 1997). One Secretariat staff member observed that, "all the money is gone to the NGOs and grassroots...donors think that governments are not doing a good job."[31] Finally, as discussed above, government delegates came to see NGOs as important partners in implementing the CCD.

Conclusions

NGOs had a high degree of influence on the process and outcome of the CCD negotiations. The link between the bottom-up approach and NGO participation in the implementation of the Convention, the relatively homogeneous and cohesive nature of the NGOs attending the meetings, and the supportive environment provided by the negotiations enhanced the ability of NGOs to exert influence on certain issues. In the absence of industrial lobbyists, who usually compete for negotiators' attention, environmental and social NGOs were able to coordinate their work and make a significant impact on the outcome of the negotiations. The setting was ripe for NGOs to become the brokers of expertise, particularly in light of the limited role of the official structure for expertise, the scientific International Panel of Experts on Desertification (Corell 1999b).

These findings supplement traditional studies of experts in international negotiations that tend to focus on the scientific advisers and emphasize the need to examine all actors with relevant knowledge. It is also useful to examine in what phases of the negotiations influence was exercised by participating actor-groups. The *pre-negotiation phase* of the CCD was from the negotiation of Chapter 12 on drought and desertification in *Agenda 21* until the organizational INCD meeting (1992–93). The *negotiation phase* lasted from INCD-1 (February 1993) until INCD-5, when the Convention text was agreed (June 1994). The *interim phase* was from INCD-6 (January 1995) until the Convention entered into force (December 1996). The *implementation phase* began in 1997.

NGOs were not present, and therefore not influential, during the pre-negotiation phase. However, during the course of the negotiation phase *and* the interim phase, they coordinated their activities and were able to act as a cohesive group. Additionally NGOs were more influential than IPED because they participated in the negotiations during the phases when NGO input on issues, such as implementation, was useful for negotiators. Nevertheless, the composition of the NGO group in the desertification case was rather unique because it lacked big northern NGOs and industry representatives. This way the high degree of NGO influence may not occur in other cases where environment and development NGOs have to compete with industry representatives, who have considerable resources and often diametrically opposed views.

There are important lessons to be learned from this analysis of the negotiations for the Desertification Convention for the study of the role of experts in international environmental negotiations. The concept of "expert," as usually employed in many studies of expertise, should be expanded to include not only scientists but also other actors who possess relevant knowledge. Issue-relevant knowledge can be provided by scientific advisers and—perhaps with greater impact—by NGOs as well. This finding is particularly relevant for the debate about global environmental problems that may be closely related to development issues such as biodiversity.

Notes

This chapter draws heavily on material that has appeared previously in Corell (1999a, b) and Corell and Betsill (2001).

1. The numbers are based on the author's calculations from lists of participants issued at the meetings. The number of accredited NGOs is often higher, and there is usually a discrepancy between this number and the actual number of NGOs attending negotiation sessions. For instance, by COP-1 a total of 360 NGOs had been accredited to the CCD process, but only 187 had attended one or more meetings. However, the participant lists may also be inaccurate in that not all participants are registered in the lists.

2. Author's notes from INCD-1, 2, and 3 (May–June 1993, September 1993, January 1994).

3. ELCI "Lobby Manual for NGOs," January 1994.

4. Interview, NGO representative 2, 27 October 1994; author's notes 19 November 1995.

5. See, for instance, International NGO Conference on Desertification, "Proposals to the Intergovernmental Committee for a Convention to Combat Desertification (INC-D)," Bamako, Mali, 16–20 August 1993 (Nairobi: KENGO, ENDA-TM, NEST, GUAMINA, 1993).

6. Interviews, INCD Chair, 29 February 1998; Secretariat staff 6, 27 May 1998.

7. Interview, Secretariat staff 6, 27 May 1998.

8. See, for example, International NGO Conference on Desertification, 1993.

9. Interview, Secretariat staff 3, 27 May 1998.

10. Author's notes, 25 January 1994.

11. Author's notes, 10 October 1997.

12. Interview, NGO representative 2, 27 January 1994.

13. Interview, IGO representative 8, 12 May 1998.

14. Interviews, desertification scientists 1 and 2, 15 May 1998; IPED member 2, 8 May 1998.

15. Interview, NGO representative 1, 26 January 1994.

16. Interview, delegate 33, 10 October 1997.

17. Interviews: NGO representative 6, 6 October 1997(A); NGO representative 7, 6 October 1997(A).

18. Interviews: NGO representative 1, 26 January 1994; NGO representative 7, 7 October 1997.

19. Interview, Secretariat staff 6, 27 May 1998.

20. Interview, Secretariat staff 3, 27 May 1998.

21. Interviews: delegate 12, 3 June 1993; delegate 17, 3 June 1993.

22. Interview, Secretariat staff 6, 27 May 1998.

23. Interview, Secretariat staff 3, 27 May 1998.

24. Interview, Secretariat staff 3, 24 January 1994.

25. Interview, Secretariat staff 6, 27 August 1998.

26. Interview, NGO representative 6, 6 October 1997a.

27. Interview, Secretariat staff 5, 27 May 1998. Another Secretariat staff member lamented the fact that the big NGOs did not attend the CCD negotiations and said that the process could use some large northern NGOs because they would have resources and could help to raise the southern NGOs' capacity. Interview, Secretariat staff 6, 27 May 1998.

28. Interview, Secretariat staff 5, 27 May 1998.

29. Interview, delegate 34, 20 February 1998.

30. Interview, NGO representative 2, 27 January 1994.

31. Interview, Secretariat staff 6, 27 May 1998.

6

Non-state Influence in the International Whaling Commission, 1970 to 2006

Steinar Andresen and Tora Skodvin

The roots to the current international regime for the regulation of whaling can be traced back to the early 1930s when the first conventions were signed. The International Convention for the Regulation of Whaling (ICRW) was set up at an international conference in Washington in 1946 and came into force in 1948. By 1950, 16 nations had ratified the convention. The International Whaling Commission (IWC) held its initial meeting in 1949.

Few international organizations have undergone more dramatic changes than the IWC (Andresen 1998). Starting out as a "whaling club," completely dominated by the short-term interests of the whaling industry, it evolved into an international regime that has maintained a moratorium on all commercial whaling for the last two decades. Currently aboriginal subsistence whaling is the only type of whaling endorsed by the majority of IWC members. The commercial whaling that is taking place is not internationally managed. The IWC has also sought to strictly limit lethal research whaling, but with modest success.

To what extent, and under which conditions, have non-state actors influenced the international regime for the regulation of whaling? Many scholars have studied the role of non-state actors in international decision-making. Most studies focus, however, on direct non-state influence at the *international* decision-making level (Betsill and Corell 2001). One main objective of this chapter is to explore non-state influence on international decision making through two main channels: directly at the international level and indirectly via the domestic channel.

In our empirical analysis we deal mainly with two major changes in IWC regulation. The first change was the adoption of a new management procedure in 1974 aiming at a stronger link between scientific assessments of whale stocks and the allocation of catch quotas. The second was the moratorium decision in 1982 (to take effect in 1985/6) that imposed a ban on commercial whaling. Both changes in policy correspond to shifts in the influence of different groups of non-state actors (Peterson 1992).[1] The first of these changes occurred as the scientific community gained increased influence on IWC policies in the late 1960s and the beginning of the 1970s. The second change took place as the environmental and animal rights movement entered the scene with full force and succeeded in mobilizing support for a moratorium on commercial whaling in the early 1980s.[2] The current phase of the IWC was introduced with the elaboration and adoption of the Revised Management Procedure during the first half of the 1990s. Although this regulatory change is not to a similar extent associated with equally significant shifts in the relative influence of different groups of non-state actors, it nevertheless indicates that the balance of influence between pro- and anti-whaling forces again may be changing. Also this tendency has become much more pronounced over time, clearly demonstrated at the last IWC meeting in 2006.

The chapter begins with a brief discussion of the term "non-state actor" within the context of the IWC and an introduction to the general background of the IWC process. The analysis is divided in two sections. In the first, we *assess the level of influence* of the two main non-state actors: the scientific community and the environmental/animal rights movement. In the second, we *explore explanations* for the observed levels of influence. We weigh the impacts of explanatory factors at both the domestic and the international decision-making levels. Empirically we distinguish between two time frames: The first runs from the late 1960s to the mid-1970s and is characterized by a quite strong scientific influence on IWC policies. The second runs from the early 1980s to the early 1990s and represents a shift in influence from the scientific community to the environmental and animal rights movement. In addition we briefly analyze the development that has taken place in the IWC since the early 1990s until present time (i.e., mid-2006). In this period the environmental and animal rights movement's dominant position has been increasingly challenged by pro-whaling forces. Our findings are summarized in a final section.

Non-state Actors in the IWC

In chapter 1 the non-state actor is defined as groups of actors that do not operate on behalf of a government or an intergovernmental organization. Within the IWC three groups of actors may fall within the category of non-state actor: the whaling industry (and more recently, proponents of the whaling "industry"), the scientific community, and the environmental and animal rights movement.[3]

The environmental movement is organized in various nongovernmental organizations (NGOs) that started to show up at IWC meetings as observers in small numbers in the mid-1960s. Although representatives of these NGOs may appear as members of national delegations in some cases, they operate independently of national governments. Even when a strong alliance exists between the environmental movement and national governments, the two constitute distinct groups.

The whaling industry, on the other hand, can be said to fall within this category to varying degrees at different phases in the history of the IWC. During the early phase (before 1960) the whaling industry dominated the scene. The industry was frequently represented in national delegations as well as being officially represented as observers (Tønnessen 1970). Even though the whaling industry developed networks through which they attempted to wield influence on IWC regulatory policies, this group's main channel of influence was through home governments. Before 1960, the majority of IWC members were engaged in whaling activities, and in many cases national interests *were* whaling industry interests. As a result, to identify the whaling industry as a distinct non-state actor during this phase may be problematic. Since the early 1990s, however, representatives of this group have reappeared as observers at IWC meetings. The new pro-whaling organizations are organized as transnational NGOs such as the World Conservation Trust and the High North Alliance.

Throughout the IWC history, science has played an important role. At the first IWC meeting in 1949 a (joint) standing Scientific and Technical

Committee was set up. While the IWC agenda and the number of working groups and subcommittees on scientific issues have vastly expanded, the basic structure of the organization still has a Scientific Committee at the very heart of its activities (Andresen 2000). Consequently, in contrast to other non-state actors, scientists have a formally institutionalized channel of influence *at the international level*.

Does this group, however, qualify as a non-state actor? Do scientists operate independently of national governments? To be considered a non-state actor, two requirements need to be satisfied. First, the community of scientists (i.e., the Scientific Committee) needs to operate independently of the Commission. Second, participating scientists need to operate independently of their national governments. Although the Scientific Committee has operated relatively independently of the Commission during the whole process, individual scientists have not always operated independently of national governments, particularly during the early phase. As further discussed below, the Scientific Committee underwent changes during the 1970s that also served to strengthen its autonomy. According to information obtained both from the Secretariat and participating scientists, the Scientific Committee enjoys considerable institutional autonomy vis-à-vis the Commission (Andresen 2000). Also the large majority of members of the Scientific Committee are merited scientists that operate independently of their appointees (national governments) in the sense that they do not operate with a political mandate (Andresen 2000: 50–51). However, relationships among individual scientists and with both the environmental movement and national governments are often strengthened during periods of strong polarization, implying a weaker scientific autonomy during such periods as well as a weaker internal unity within the scientific community (see also Schweder 2000, 2001). Nevertheless, on this basis we conclude that scientists seem largely to operate in their capacity as scientists, not as representatives of the governments or organizations that have nominated them, and that the community of scientists thus constitutes a non-state actor during the phase of the IWC process under scrutiny here.

In different time frames of the process these groups of non-state actors have represented competing interests to varying degrees (Peterson 1992).

For one thing, as noted above, the groups have neither been active during the whole process nor necessarily at the same time. Indeed only scientists have been active from the early phases of the process until now. Whalers were active from the beginning until the late 1960s, when so few whalers remained that there were only remnants left of the Japanese and the Soviet whaling fleets. At the time of the adoption of the moratorium in 1982 only Japan was doing industrial whaling. Although environmental groups started showing up as observers at IWC meetings in the 1960s, they can only be considered an active non-state force in the IWC as of the early 1970s. In this analysis, our main focus is on scientists and environmentalists, but in our discussion of the current situation, we also take the new pro-whaling organizations into account as a counterbalancing force that increasingly makes its mark on the IWC process.

Initially scientists and environmentalists shared the same concern: the rate at which whales were harvested presented a serious threat to their survival and, as such, greater restrictions on whaling were necessary. Contentions between these two groups and *within* the scientific community started to show during the late 1970s and became explicit with the moratorium decision at the beginning of the 1980s. The environmentalists pushed for a full moratorium on all species, but the majority of scientists argued that this was unnecessary and not scientifically warranted. Concurrently a strong minority within the scientific community supported the moratorium decision. During the 1980s, however, the two groups increasingly represented different opinions. The majority of scientists in the Scientific Committee argued that new and improved knowledge indicated abundance of certain species (most notably minke whales). Environmentalists either did not accept the scientific estimates of whale stocks (i.e., they argued that the estimates were more uncertain than scientists acknowledged and/or too uncertain to warrant commercial whaling) or opposed whaling more generally for ethical or political reasons. Currently the differences of opinion between these two groups have become less clear-cut as the environmentalists are more divided than they used to be, but there is no reason to assume any significant convergence, at least not in the short term. Even today the two groups of actors can be seen as proponents of different views.

Assessment of Non-state Actor Influence in the IWC: 1970 to 2006

The International Convention for the Regulation of Whaling (ICRW) was adopted in 1946. The IWC was set up two years later, and its initial meeting was held in 1949. Although the Commission was open to all, in the early years only some 15 members participated, most members had some connection to whaling. From the Preamble, the purpose of the Whaling Convention is known to be to conserve the whales in order to secure the orderly development of the whaling industry. This objective was considered novel at the time, as it attempted to strike a balance between conservation and utilization. From this official goal, it is clear that conservation was seen mainly as a means for securing orderly utilization. Detailed catch regulations are adopted in a Schedule, which is an integral part of the Convention. Changes to the Schedule are made by a three-quarters majority. In the early years there were no national quotas, only one total quota existed. This was open to all, meaning that all could compete to catch as much of the total quota as possible. The Convention states that all decisions are to be based on "the best scientific advice," thereby emphasizing a dependence on science. Indeed, a Scientific Committee was established at the first meeting of the IWC in 1949, where all states could send representatives.

The history of the IWC can be divided into distinct phases. The first phase runs from the establishment of the IWC until the early 1960s. These years were characterized by *overexploitation* and commercial depletion of whales. The second phase covers the 1970s, after a transitional period during the 1960s, and over this decade more *balanced management* of the whaling resource was attempted. The third phase, which mainly covers the 1980s, is characterized by the *protection* of whales. A fourth phase seems to have emerged since the mid-1990s characterized by less support for the ban on commercial whaling. In this chapter we concentrate our attention on the second and third phases, but developments during the fourth phase are briefly analyzed. To illustrate the significant changes in the IWC, however, we take a very brief look at the first two decades of the organization's history.

During the initial period of the IWC, few countries sent representatives to the Scientific Committee, and the state of knowledge was limited and disputed (Schweder 2000). Moreover the environmental movement was absent from the scene, so industry was to a large extent the main player and provided the decision premises for state members. The relevance of scientific input was recognized, given the decision to establish the Scientific Committee, but scientific warnings of overexploitation were disregarded.

This state of affairs started to change during the 1960s. The gradual emergence of a "new" IWC was, at least partly, science driven. Around 1960 there was a real risk that the IWC could break up over disagreements on quotas and catch limits, and some members left the IWC for this reason.[4] On the initiative of the United Kingdom, a Committee of three independent scientists (later four) was established. This committee succeeded where the Scientific Committee had failed; it quantified the necessary catch reductions needed. The regulations worked for a few years and their conclusions were generally accepted by IWC members. Thus toward the late 1960s, the IWC followed the advice of the Scientific Committee, and in 1967, catches were finally within sustainable limits (Scarff 1977). This brings us to the period under study in this paper: 1970 to 2006.

Our empirical analysis explores two questions: To what extent did non-state actors influence IWC regulatory policies in the period from 1970 to 2006? Which non-state actors acquired influence and why? We are thus concerned with (1) the influences of non-state actors on IWC regulatory policies and (2) the relationship between the two main groups of non-state actors that were active in this phase of the IWC process: the scientific community and the environmental movement.

1970s: New Management Procedure

During the 1970s the influence of the scientific community remained fairly high, although it decreased toward the end of the period. This does not mean, however, that scientific advice was automatically followed. There were discrepancies and time lags, but overall, the match between advice and regulations was increasing. A number of new regulations were adopted, implying a much more cautious management of the resource. For instance, the arbitrary "blue whale unit"[5] was abolished, new species were included in the management repertoire, and those species most at risk of extinction were completely protected (Andresen 2000).

To secure more cautious management, a new management procedure was suggested by Australia, and this procedure can be seen as a compromise proposal since the anti-whaling countries were not yet able to get a moratorium adopted (Bailey 2006). The 1974 procedure proved hard to implement because of lack of data. Nevertheless, while it did not immediately lead to more restrictive regulations, it contributed to raise the level of scientific argumentation that went into decision making (Peterson 1992). Previously, "the members of the Scientific Committee had given the IWC a unanimous 'best estimate' resting as often on political as scientific grounds without giving any explicit account of the criteria actually used in making the estimate" (Peterson 1992: 164). In response to external pressures, inter alia, the Scientific Committee used the adoption of the new procedure to establish a more open process "in which papers were published, commentary was sought, and the scientific basis of the conclusions was made explicit" (Peterson 1992: 166). Most important perhaps, the new procedure reinforced the significance of the Scientific Committee itself, since it mandated far more data and more accurate models of whale population dynamics. Thus the number of scientists in the Scientific Committee increased, the workload expanded considerably, and ever larger numbers of invited independent scientists participated in the Scientific Committee meetings (Andresen 2000). The development during the 1970s therefore went in the direction of stronger scientific impact on IWC regulatory policy. In this development a new procedure and institutional arrangements in turn served to further reinforce the scientific contribution to the process.

The change that occurred in the early 1970s in the IWC to some extent was driven also by other factors besides science. In particular, the full picture should include the state of whale stocks. By the 1970s some whale stocks were so depleted that most countries had lost interest in large-scale whaling in the Antarctic. Thus the profitability in industrial whaling was reduced, not because of stricter regulatory policies, but because there simply were not enough whales left to catch. During the 1960s the level of catches no longer kept pace with the quotas (Peterson 1992). Whaling nations only reduced their catch when they no longer were able to fill their quotas. Thus, while the impact of science was increasing during this period, it was not the only reason for the more cautious management policies adopted by the IWC (Andresen 2000). Moreover discrepancies and time lags occurred between advice and (changes in) regulatory policies. We thus assess the influence of the scientific community to be at a *moderate level* according to the criteria introduced in chapter 2.

1980s: Moratorium
In the early 1970s, as scientists were starting to expand their newly acquired influence on IWC policy by way of the new management procedure (Peterson 1992), another shift took place in the IWC. The scientists were overtaken by the environmental movement. This was quite a mixed bag of organizations. The first on the whaling scene were animal rights groups such as the International Society for the Protection of Animals and the Fauna Protection Society (FPS). The first environmentalist group to send observers was Worldwide Fund for Nature (WWF) (1965), and at the time of the 1972 Stockholm Conference the whaling issue was coined primarily in environmental terms (Bailey 2006). NGO attendance at the IWC increased strongly from a handful in 1972 to 57 in 1982 (Andresen 1998). Traditional green NGOs like WWF and Friends of the Earth were regulars, but more "exotic" groups like the "Assembly of Rabbis" and "A&M Records" also regularly showed up. Greenpeace was a latecomer, attending since 1978, but soon turned out to be the most important, influential and aggressive anti-whaler NGO.

Towards the end of the 1970s Greenpeace increasingly made its mark on the process. To the modest extent that the environmental movement took part in the whaling debate in the 1960s, their arguments were largely in line with the arguments of the scientific community. By the late 1970s/early 1980s this was no longer the case. In contrast to the scientific community, the environmental and animal rights organizations, albeit on varying grounds, lobbied strongly and vocally for a moratorium on whaling.

The environmentalists' demand for a ten-year moratorium on commercial whaling was successfully put forth as early as 1972, when the UN Conference on the Human Environment (the Stockholm Conference) unanimously adopted a recommendation to that effect. One main reason why it took ten years before this demand was accepted in the IWC lies in differences in the composition of participants at the two arenas. At the time the IWC was primarily composed of current or previous whaling nations. The large influx of new members that eventually would shift the balance toward a moratorium decision had not yet arrived. Although the environmental movement had an important ally in the United States (on both arenas), most IWC members did not support the call for a moratorium.

In 1982, however, the environmental movement achieved their goal when the IWC, with a three-quarters majority, adopted a moratorium on commercial whaling to take effect in 1985/6. The moratorium called for a stop in commercial whaling, pending a "comprehensive review" of all whale stocks to be conducted no later than 1990. Today, twenty years later, the moratorium still stands. Four countries reserved themselves against the moratorium, but only Norway upheld its reservation, and Norway is currently the only IWC member that conducts commercial whaling. Japan and Iceland conduct scientific whaling, and there are also some aboriginal whaling nations, including the United States.

The increased influence of the environmental movement seems to have been gained at the expense of scientific influence. That is, from the late 1970s and through the 1980s the influence of the environmental movement increased while the influence of the scientific community (notably the Scientific Committee) decreased accordingly. Preceding the moratorium decision, the Scientific Committee claimed that a full moratorium was not scientifically warranted and argued that a more nuanced approach was needed. It should further be noted that there was no scientific consensus on this point. A significant, able, and vocal minority of the Scientific Committee claimed that a moratorium was exactly what was needed in order to get improved knowledge of whale stocks and whale population dynamics (Andresen 2000).

In sum, in the late 1970s/early 1980s influence in the IWC shifted from scientists and the Scientific Committee in favor of the (relatively) new non-state actor on the scene, the environmental movement. By the criteria introduced in chapter 2, the influence level of this group can be assessed to be *high*. The environmental movement succeeded in shaping the negotiation process and, by exerting considerable influence on the framing of the whaling issue, also in shaping the negotiation outcome. This was done by coaching representatives of the large majority of the new member states and thereby securing the three-fourths majority necessary to get the moratorium adopted.

As noted, the environmental movement had an important ally in the US government. The United States had stopped all commercial whaling operations by 1970, the same year that the environmentalists started taking an interest in the whaling issue. Since the early 1970s the United States can be argued to have been the single most important actor on the whaling scene. "Though entitled only to one vote in the IWC, it had what the IWC and any other single government or group of governments lacked: the ability and the will to enforce restrictions against others by invoking the trade sanction provisions of various domestic laws" (Peterson 1992: 172). Thus, given the strong overlap in the interests and positions of the environmental movement and the US government, it is difficult to determine which of these factors actually caused the change in IWC policy.[6] However, as pointed out by Peterson (1992), a conclusion to the effect that it was US policy alone that brought about the change in the IWC begs the question of why it took them so long to use its capability. We take this aspect as strong indication that the environmental movement's activities had a direct and strong impact on the adoption of the moratorium through active recruitment of new anti-whaling members as well as very effective lobbying—something we will have more to say about when explaining this development. Its alliance with the United States, rather than representing an alternative explanation to *IWC policies*, constitutes a supplementary explanation to *environmentalist influence*, particularly in the sense that the environmental movement's arguments served to legitimize the US position on whaling.

1990s: A New Phase?
The picture has not changed dramatically since the early 1990s in the sense that there is little chance that the moratorium will be lifted anytime soon. Nevertheless, there are strong indications that the dynamics of the IWC process again are changing in a direction that have implications for

the balance of influence between the different non-state actors involved in the process. Since the early to mid-1990s there have been some indications that the environmental movement is not as influential as it had been. First, there has been a rather significant increase in the catch by IWC members during this period. Norway resumed commercial whaling in 1993 and the catch quota has increased over time. For the 2006 season the Norwegian quota is more than a thousand minke whales in the northeast Atlantic. Similarly Japan's scientific whaling has increased strongly, and it has also been expanded to include new species. In 2003 Iceland rejoined the IWC after having left the organization in frustration in 1991 but has resumed scientific whaling. Aboriginal whaling is on the rise. In short, the number of whales killed is increasing rather rapidly despite the moratorium (Black 2005). As many as 24,314 whales have been killed since the adoption of the moratorium of the mid-1980s (Bailey 2006).

Second, there has been a strong recruitment of new members by both sides. While the United States and Greenpeace were most successful on this account in the 1970s and 1980s (see below), this time it appears that the pro-whaling side, particularly Japan, has been more effective. Environmental NGOs have long claimed that Japan has used development aid as a means to secure support for their view in the IWC, and at the last IWC meeting (2006) there was a small minority (33–32) in favor of whaling. Although this is not even close to lifting the moratorium, the direction of the IWC will change as a result of this development.

Third, although delayed for three years, the Revised Management Procedure developed by the Scientific Committee was adopted by the IWC in 1994. The Revised Procedure opens up for commercial whaling and thus stands in stark contrast to the environmentalist position. Delaying tactics have been applied, however, on the insistence of the (previous) majority of the IWC that a Revised Management Scheme, detailing inspection and observer systems, has to be in place before quotas can be considered. This may change if the pro-whaling majority holds, but it will not become operative before the moratorium is lifted.

This development indicates that, at least in a long-term perspective, a shift in the influence balance between scientists and the environmental movement is in progress. This shift is further supported by the reappearance and increasingly marked role of the third non-state actor in this process, the new pro-whaling organizations whose position to some extent is based on, and legitimized by, the recommendations of the Scientific Committee. This trend, however, should not be exaggerated as the end of the moratorium is nowhere in sight. The actual impact of this development still remains to be seen.

Exploring Explanations

Scientists gained increased influence from the early to mid-1970s, reaching a moderate level of influence in the development of a new management procedure. The environmental movement was most influential in the 1980s when the IWC adopted a moratorium on commercial whaling, which was the primary target for this group. These two groups represented at least partially competing forces during the 1980s, and the increase in influence that the environmentalists experienced seems to represent a loss in influence for the scientific community. Since the mid-1990s, scientists seem to have regained some authority at the expense of environmentalists, and the trend seems to have resulted in a slight shift in favor of scientific opinion.

In this section we explore the question of why these groups gained or lost influence on the process and which factors seem to have caused the shift in influence between them. The following represents an *exploration* into possible explanatory factors rather than a fully developed, in-depth and methodologically stringent empirical analysis.

1970s: Scientific Dominance

Four factors seem important for understanding the level of scientific influence in the IWC during the early to mid-1970s. First, there was an increasing demand for advanced knowledge on stocks and population dynamics, and the quality of the knowledge input was improved significantly as compared to the previous phase. Second, the knowledge generated was associated with a stronger scientific consensus than previously. Third, the scientific body of the regime was strengthened particularly in terms of ensuring a higher level of independence. Fourth, counterbalancing forces were considerably weak.

Increasing Demand and Improved Quality   As noted, by the late 1960s and early 1970s whale stocks were depleted to the extent that whalers no longer managed to fill their quotas and the profitability of (Antarctic) whaling was in sharp decline. Even whaling managers acknowledged the need to adopt a more science-based approach to the regulation of whaling (Peterson 1992). During this period scientists were in command of a resource—advanced knowledge on whale stocks—for which there was an increasing political demand. This demand was directed to the scientific body within the IWC organizational structure—the Scientific Committee—it was not directed to individual scientists at the national level. In the early 1970s the Scientific Committee thus had the initiative and could, at least to some extent, set the premises for the debate that took place within the IWC. One implication of this development seems to be that the main arena for non-state influence during this period was at the international level, particularly linked to the work of the Scientific Committee, and we have no indication that domestic channels of influence were of particular significance for the influence of non-state actors.

As the profitability of Antarctic whaling declined, the importance of whaling in other areas rose, and the IWC started using increasingly precise definitions of stocks. This trend resulted in the adoption of the new management procedure in 1974 (Peterson 1992). The new procedure mandated more, and more *accurate*, data. Notably the new procedure implied a more precise differentiation between stocks in terms of depletion risk. Whale stocks were divided into three classes: initial management stocks, sustained management stocks, and protected stocks. This classification was based on a comparison of current stock population size to the size that would supply the maximum sustainable yield. Quotas were then allocated in accordance with the classification of the whale stock in question (Andresen 2000; Peterson 1992). This differentiation proved very difficult to implement, however, which also was one reason why some scientists later argued in favor of a full moratorium.

Scientific Consensus and More Independence   The fact that the Scientific Committee managed to adopt a new procedure indicates a relatively high level of scientific consensus, which served to increase the weight of scientists' advice. This was in sharp contrast to the preceding period.

Peterson (1992: 160) maintains that until the late 1950s, the case for more restrictive whaling management procedures "suffered from the cetologists' inability to present detailed consensual advice or compelling arguments that uncertainty should always be resolved by erring in the more restrictive direction."[7]

The period was characterized by a strong institutional buildup, not the least with regard to the scientific component (Andresen 2000). In 1974, the IWC got its own secretariat. In the same year the United States proposed that observers from the Food and Agricultural Organization and the United Nations Environment Programme be allowed to participate in Scientific Committee discussions. This was accepted upon the discretion of the Chair, and in 1977, scientifically qualified observers were permitted to attend the Scientific Committee for the first time. Since then they have regularly availed themselves of this right.

During the early phase of the Scientific Committee relatively few scientists attended, and they often had very close ties to their national governments (Andresen 2000). The new procedure paved the way for a permanent presence of scientists that are not aligned with any particular nation. This scheme has worked to remedy the problem of tacit or open pressures for scientists to conform to national preferences. By increasing the number of invited scientists and including scientists from other intergovernmental bodies, the basis and independence of the scientific input generated in the Scientific Committee has broadened and increased. Besides serving to reinforce the significance of a formalized channel for scientific influence at the international level, the broader participation of scientists seems to have reduced the polarization within the IWC (Andresen 2000). With polarization it seems there is, in general, less likelihood of scientific influence on international decision making (e.g., see Underdal 1989).

Interestingly a key actor in the process of strengthening the scientific component of the IWC was the United States. The United States had ended its commercial whaling activities in 1970 and adopted a more conservationist view toward whaling management. As a more conservationist element was introduced in the IWC, the scientific component was strengthened because scientists had long argued, since the 1950s, for a stronger emphasis on conservation for IWC's purpose. While the shift in

American domestic whaling policies took place in 1970, the shift in US policy toward the employment of more coercive policy instruments at the international level did not happen until somewhat later.

Weak Counterbalancing Forces   There was no real competition for influence among non-state actors during the 1970s. The environmental movement had yet to start mobilizing on this issue, so it generally supported the position of the scientific community. The only "opponent" to the scientific community was the remaining whaling industry, represented by Japan and the Soviet Union.

Even with a gradually increasing influence over IWC policies, scientists were still frustrated by the slow pace at which changes took place, so they tried to develop additional channels of influence (Peterson 1992). The Food and Agricultural Organization provided the most prominent channel. It made continued cooperation on whale stock assessment conditional upon the adoption of policies that more closely reflected the growing scientific consensus on quotas. The scientific community also attempted to activate a domestic channel of influence by calling public attention to the issue, but the public interest necessary for this strategy to succeed did not yet exist. By the time that it did, the scientists were overtaken by the environmental movement (Peterson 1992). An interesting footnote can be added to this development. J. L. McHugh, the US Commissioner in 1965, held onto the opinion that the best strategy to get the IWC on the right track was to inform the public on the issue. This was a strategy he later regretted as he described the ignorance of the public and the forces they had unleashed that were now out of control—like the "sorcerer's apprentice" (McHugh 1974: 313).

1980s: Environmental "Capture"

Four factors seem particularly important for understanding the shift in influence from scientists to environmentalists in the late 1970s/early 1980s. First, the environmental movement successfully mobilized the general public on the whaling issue and thus contributed significantly to a stronger public concern over the state of whale stocks. Second, an increased public concern enhanced the importance of domestic channels of influence. The environmental movement seems to have had a strong impact on national positions on the whaling issue in key IWC member countries and the US government in particular. Third, in addition to powerful political resources, the environmental movement had financial resources that they seem to have used in rather untraditional, but powerful, ways to influence IWC policy. Fourth, counterbalancing forces were weak.

Increased Public Concern   By the late 1970s/early 1980s, the concern and sense of urgency about overexploitation had spread beyond whalers, cetologists, and IWC member countries to the wider public in Western (and eventually also non-Western) countries. Overexploitation of the whale stock is perhaps the single most important factor to explain the increased influence of the environmental movement during this period (Peterson 1992). The development of public interest in the whaling issue was not the least due to the activities and mobilization campaigns orchestrated by the environmental movement.

For mobilizing public opinion, the environmental movement has a broader and more powerful set of means at its disposal than scientific communities. Because the scientific community gains its legitimacy from the provision of objective and policy-neutral information, its active mobilization for a specific (political) position on an issue can backfire and jeopardize the very basis upon which its influence rests.

In contrast, one important means of public mobilization employed by the environmental movement in the whaling process was the creation of the whale as a symbol—a "super-whale." Kalland (1993, 126) notes:

[W]e are told that the whale is the largest animal on earth (this applies to the blue whale), that the whale has the largest brain on earth (the sperm whale), that the whale has a large brain to body weight ratio (the bottlenose dolphin), that the whale has a pleasant and varied song (the humpback), that the whale is friendly (the gray whale)...and so on. By talking about *the* whale, an image of a single whale possessing all of these traits emerges. But such a creature does not exist. It is a mythic creation—a "super-whale," which has come to represent all species of cetaceans.

Domestic Channels of Influence and Powerful Allies   Rising public interest in the issue contributed to an increased importance of domestic channels of non-state influence and at that arena the environmental movement seems to have a marked advantage over the scientific community. It is easier to mobilize the general public on a slogan like "Save the whale" than it is to mobilize the public with scientific statements on the conditions and procedures under which whales can be harvested in a sustainable manner. A stronger public concern over the whaling issue and the enhanced significance of domestic channels of influence that followed from this development seem to constitute primary sources of the environmental movement's enhanced influence on IWC policies.

A key actor in this regard was the United States. Environmental concerns and particularly the protection of (perceived) endangered species reached the agenda very early in the United States. The passage of the Endangered Species Act (ESA) in 1966 was an important step in this process. Environmental groups played a key role in this process. Their pressure in 1969 resulted in a revision of the ESA and creation of a list of endangered animals worldwide (Bailey 2006: 21). Hearings before Congress in 1971 led to the US adoption, as national policy, of a moratorium on all species of commercial whaling. The United States therefore became the first government to quit commercial whaling for environmental reasons, not the least due to pressure from environmental groups. Congressional hearings showed that animal welfare organizations were in the forefront, but the cause was soon embraced by the environmental movement (Bailey 2006). The environmental movement thereby had a strong ally in the US public, Congress, and the Nixon administration. The environmental movement in the United States was in command of powerful political resources on this issue, and it had a considerable impact on American whaling policies, again not the least due to active public mobilization (Freeman 1994; Kalland 1994). Bailey (2006: 26) argues, "By 1978 it would seem safe to say that the anti-whaling 'norm' had been internalised in the US. Even Walter Cronkite, 'The Most Trusted Man in America' publicly sided with those who 'believe it is morally wrong to kill cetaceans' except perhaps for reasons of subsistence.''

As a result of public pressure the US government's support for the Scientific Committee's conservationist approach was rapidly transferred to a protectionist stance on commercial whaling in the wake of the 1972 Stockholm resolution.[8] The increased public attention was also reflected in a steady increase in the number of US delegates throughout the 1970s and anti-whaling organizations were always represented on the delegation.

The US therefore was a key actor in the IWC because of its ability and (increasing) willingness to use sanctions against nations that did not comply with its position on the whaling issue. US legislation, notably the Pelly amendment to the Fisherman's Protective Act, empowered the US Secretary of Commerce to certify a nation that is acting in a manner that diminishes the effectiveness of a multilateral agreement to which the United States is a signatory party (see DeSombre 2001). The United States used this legislation for two main purposes: to bring into the agreement nations that were whaling but were not members and to bring about changes in the whaling policies of nations within the agreement. There is no doubt that the main reason why the number of whaling nations was brought down from twelve to zero from 1985 to 1988 was US power politics (Andresen 1998). The United States, for instance, threatened Japan, Iceland, and Norway with economic sanctions if they did not change their whaling policies in accordance with the US position, and only Norway stood up to US threats (Andresen 2004).

The environmental movement made use of its transnational network to coordinate activities and influence national positions on whaling on a broader scale. Most of the green NGOs like WWF and Friends of the Earth used the traditional means of persuasion and "shaming and blaming." Greenpeace, however, also applied other means. In combination with the US government's threat to enforce economic sanctions against whaling nations, Greenpeace threatened to enforce boycott actions against these same nations. For example, after the moratorium decision, Greenpeace organized a boycott of Norwegian fish products and the US government threatened with economic sanctions. In 1986 Norway decided to halt its commercial whaling operations. Iceland faced similar pressures from Greenpeace and the US government and stopped its research whaling in 1989. Although disputed, Iceland's loss of revenue as a result of the Greenpeace boycott actions is estimated at USD 30 million (Andresen 1998). Some so-called green activists, like the Sea Shepherd Conservation Society (led by Paul Watson) resorted to violent measures such as sinking a Norwegian whaling boat. The whaleboat sinking did backfire, however; it bolstered the pro-whaling sentiment in Norway. Nevertheless, the environmental movement, through a wide variety of means (together with the US government) succeeded in influencing the positions of whaling nations. Although the material interests of the target nations were limited, consession to the strong anti-whaling norm became an opportunity in this issue-area for nations to get "green" at no economic costs.

Financial Resources and Untraditional but Powerful "Instruments of Persuasion" The responsiveness of the general public to the environmental movement's campaigns on whaling increased income for environmental organizations. Andresen (1998: 441) notes, "it seems fairly safe to assume that some of the major NGOs like Greenpeace have profited greatly in the form of higher contributions resulting from public concern about whaling." While contested and hard to document, allegations have been repeatedly set forth that the environmental movement used at least part of their newly acquired financial resources to influence IWC policies in rather untraditional ways. In concert with the US government, the environmental movement, in particular, Greenpeace, became instrumental in actively recruiting new nations as members, thereby generating a new, anti-whaling majority in the IWC.

The ICRW is open to all nations regardless of their substantive interests in whaling and it adopts changes to the Schedule by qualified majority voting. Beginning in the late 1970s, participation in the IWC skyrocketed after the US government succeeded in opening all sessions to NGO observers, and it appears that the environmental movement's role in this development was in "buying" new nations into the agreement (Andresen 1998). According to DeSombre (2001: 187), "the IWC secretary tells the story of an unnamed member state that simply signed over the check from an environmental organization to pay its dues." Similarly "a former Greenpeace consultant tells of a plan that added at least six new anti-whaling members from 1978 to 1982 through the paying of annual dues, drafting of membership documents, naming of a commissioner to represent these countries, at an annual cost of more than USD 150,000" (DeSombre 2001: 187).[9]

These allegations are controversial and disputed. Whether or not the environmental movement paid the dues of the new members, the recruitment of a new anti-whaling majority within the IWC seems to have been a strategy employed by both the environmental movement and anti-whaling nations like the United States.[10] Observers have maintained that "Greenpeace had a deliberate strategy to 'pack the IWC' with new non-whaling members...." (Andresen 1998: 439). The result was that the majority of the IWC shifted in favor of a moratorium on commercial whaling. A former legislative director of Greenpeace's Ocean Ecology Division claims that "with startling speed [environmental and animal welfare groups] carried out what amounted to a coup d'etat in the IWC" (Andresen 1998: 440). As noted earlier, recently Japan seems to have adopted the same strategy by recruiting new pro-whaling members, not only by persuasion. Paradoxically, the green movement now accuses Japan of the same tactics as the pro-whaling forces used to criticize the green NGOs.

Weak Counterbalancing Forces   The wide arsenal of instruments employed by the environmental movement in the 1980s was not available to the scientific community. The scientific community was not in command of other, equally powerful instruments. In addition the increasing polarization between pro- and anti-whaling members of the IWC was reproduced within the scientific community. While the majority of scientists maintained that a full moratorium was not scientifically warranted, a vocal minority supported the moratorium decision, thereby splitting the scientific community (Schweder 2001). Among the many factors that contributed to the contentions characterizing the Scientific Committee in this period was the demand for an increasingly accurate and broad knowledge base. This meant that new disciplines had to be brought into the Committee's work, in particular, marine biologists and statisticians. Andresen (2000) maintains that with this inclusion a communication deadlock occurred among the disciplines within the Committee, and during a period when transparency, and in particular, polarization was high and increasing, the Committee was unable to generate consensual scientific advice. Also the efforts to strengthen the Scientific Committee by broadening the basis of its conclusions during the previous period seems to have "backfired"—at least temporarily—making it more difficult to generate consensual scientific advice.

To sum up, it appears that *implementation* of the moratorium was achieved through a wide arsenal of more or less coercive means. The United States and NGOs like Greenpeace were crucial partners on this account. In the *adoption of the moratorium*, the IWC made active use of voting and recruited new member nations whose positions they influenced. In the anti-whaling movement the environmental community was the main actor.

1990s: Environmentalists on the Defensive?

Developments in the IWC indicate that in the 1990s the Scientific Committee regained some of its authority from the 1970s at the expense of the environmental movement. In this section we briefly explore the role of two factors in this development. First, paralleling the development in the 1970s, the quality of the knowledge improved because of stronger scientific consensus. Second, as became very evident during the 2006 IWC meeting, counterbalancing forces regained strength with a corresponding weakening of the environmental movement.

Improved Quality of Scientific Input and Stronger Scientific Consensus

An important aspect in the adoption of the moratorium in 1982 was scientific uncertainty with regard to the status of key whale stocks. The decision was made moreover on the condition that a comprehensive scientific assessment of potentially exploitable whale stocks was carried out. As part of this assessment, scientists also began to revise and improve the (new) management procedure from 1974. In 1991 the Revised Management Procedure was adopted by the Scientific Committee. The process revealed that with the application of the revised procedure, certain whale stocks could be harvested commercially without any danger of depletion. Ray Gambell (1995, 710), IWC Secretary for more than 30 years (until 2001), stated: "The procedure is very conservative compared to anything that has gone before, and also by comparison with management regimes for other wildlife and fisheries resources."

More certain, more advanced, and, not least, more consensual scientific knowledge seems to be a key factor for the shift in the influence balance between anti- and pro-whaling forces. There are very few serious actors today who dispute that careful commercial whaling can be undertaken in a sustainable manner. For example, even the United States recognizes that the moratorium is not supported by science and now "justifies its position as driven by public opinion" (Bailey 2006: 27). Previous whaling nations (notably, Japan, Norway, and Iceland) played a key role in the provision of new and more precise knowledge on the status of whale stocks. In the late 1980s and early 1990s they launched large-scale scientific programs within the framework of the Scientific Committee. They were keen to demonstrate that sustainable commercial whaling was possible. In the course of the process, however, they also had to admit that some of the population data of previous assessments in fact had been weak.

In Norway, the strong scientific consensus was important for its development of a domestic political consensus on the decision to resume commercial whaling in 1993. Gro Harlem Brundtland's decision to stop commercial whaling in 1986 was based on her impression that—scientifically—Norway had a weak case. The Scientific Committee's conclusion that the Northeast Atlantic minke whale could be harvested in a sustainable manner was a crucial element in Norway's decision to resume whaling (Andresen 2004). Through similar research projects Japan and Iceland demonstrated that the stocks they were interested in could sustain modest harvest. They also got support for their estimates in the Scientific Committee, but they feared sanctions from the United States and Greenpeace and therefore chose not to resume commercial whaling (Friedheim 1996, 2001).

Finally, it seems that the prominence of the whaling issue on the international agenda, and particularly, the anti-whaling norm, was gradually reduced and weakened during the late 1990s and especially since the turn of the century. This is indicated, for instance, by the response Norway's whaling policies generated. Norway's 1993 decision to resume commercial whaling in line with the very cautious Revised Management Procedure was met with boycott threats from Greenpeace and certification for sanctions by the United States. In contrast, Norway's 2001 decision to resume export of whale products passed almost unnoticed. In 1999 and 2001, a majority of some 150 members of the Convention on International Trade in Endangered Species (CITES) supported the Norwegian proposal to down-list the North Atlantic minke whales from the list of endangered species, but the required number of votes was not reached to get it down-listed (Andresen 2004). The reduced prominence of this issue is also reflected in the "Earth Summits" that have taken place since the 1972 Stockholm Conference, where the question of a moratorium on commercial whaling first was raised. At the United Nations Conference on Environment and Development 20 years later (in Rio 1992), sustainability, not protection, was emphasized. In Johannesburg in 2002 the whaling issue did not even surface on the agenda of the open meetings.

The increased authority and influence of the scientific community has not resulted in changes in regulatory policies. However, the improved knowledge has no doubt contributed to serve as an important legitimizing function for the strengthening of the pro-whaling forces.

Stronger Counterbalancing Forces   The pro-whaling NGO group has grown considerably since the 1980s when it was virtually nonexistent. The only actor on the scene was the World Council of Indigenous People (since 1988). From 1992 they were joined by groups like the High North Alliance, Friends of Whales, and thereafter an increasing number of organizations like the Group to Preserve Whale Dietary Culture. The anti-whaling sentiment emanated from countries, such as the United States, with strong NGO and lobbying traditions. The previous whaling nations did not share these traditions, and NGOs based in these countries entered the stage much later. Subsequently pro-whaling groups understood, however, that lobbying and campaigning were essential elements in the "battle over the whales." When the pro-whaling side understood the significance of NGO lobbying, both Japan and Norway became active supporters of the new pro-whaling organizations (primarily with financial support). Some of these organizations, like the World Conservation Trust and the High North Alliance, are quite influential and innovative.[11]

Domestically this new group of NGOs has become increasingly active. Together with local communities in previous whaling areas, the High North Alliance lobbied actively, and successfully, to get Norway to resume commercial whaling. In the early 1990s, moreover, Norway paid professional groups in the United States to lobby for the Norwegian position in Congress (Andresen 1998). Subsequently other pro-whaling groups have also lobbied in Congress.

Concurrently with the gradual strengthening of the pro-whaling NGO community, there are signs of a parallel weakening of the environmental movement on this issue. The strong anti-whaling community is, for instance, not as united as it used to be. While the animal rights movement and Greenpeace, not surprisingly, still oppose commercial whaling, WWF has modified its position. At the end of the 1990s, WWF declared that commercial whaling could be accepted, although on a very limited scale.[12] A main reason for this shift was fear that the IWC would break up if this was not accepted. In April 1992, the North Atlantic Marine Mammal Commission (NAMMCO) was established, with Norway, Iceland, Greenland, and the Faroe Islands as member states. Although NAMMCO in its present form does not represent a threat to the IWC, it has been successfully used as a bargaining chip by pro-whaling countries.

In Norway most environmental NGOs were opposed to commercial whaling during the 1980s. This is no longer the case. With the exception of Greenpeace-Norway, most of them accept commercial whaling that is conducted in a sustainable manner. This was important in the pro-whaling organizations' lobbying campaign for Norway's resumption of commercial whaling. These organizations acquired access to the highest decision-making levels, and they were listened to, not least because there was no active lobbying by environmentalists on the domestic level.[13]

The anti-whaling forces are still strong—and strong enough to prevent a lifting of the moratorium on commercial whaling for the foreseeable future. This discussion nevertheless shows that the balance of influence between the different non-state actors is continually shifting.

Conclusion

While studies of non-state influence often focus on influence indicators linked to the international level, this analysis indicates that the domestic level is of equal, if not greater, significance. For example, the scientific community and the environmental movement had equal opportunities to provide written information at the international level, to provide advice, and to shape the agenda. Because of the scientific community's access to a formalized channel of communication at the international level, however, this group did have a somewhat higher ability to incorporate text in the agreement. Even if the scientific community had equal (or higher) access, opportunity, and ability at the international level to provide decision premises for the debate that took place in the IWC, the international level was not the most important decision-making level for non-state influence during the 1980s. The environmental movement could utilize a very powerful channel of influence at the domestic level that gave them a much higher influence on the process than the scientific community during this phase.

A second general conclusion from this study is that the single most important determinant of scientific impact is the scientific community's ability to generate consensual advice. This is not a novel conclusion. Most studies of the relationship between science and politics have drawn the same conclusion (e.g., see Underdal 1989). Scientific consensus does not necessarily generate political consensus. It nevertheless seems to constitute a necessary condition for scientific impact. While consensus and unity may be assumed to be important for other non-state actors as well, it seems that these groups have a broader set of instruments both to acquire influence and to maintain unity. Scientific consensus seems more vulnerable to polarization, and when polarization is high, the mechanisms through which scientific consensus is established and maintained break down.

A third general conclusion is the significance of how an issue is framed for the turn of events and the subsequent influence of various types of non-state actors. From being a debate over the size and sustainability of whale resources, the whaling issue increasingly took the form of a debate on the ethics and morality of whaling more generally. This new discourse was the true-born child of the environmental movement, turning the issue into what is "right" and "wrong" rather than what is scientifically justified. As we have seen, increased influence over the framing of the issue also generated increased influence over outcomes. When an issue turns into a debate over values, scientific input has little to offer and its influence will be reduced accordingly.

The environmental movement has been instrumental in maintaining a moratorium on commercial whaling that has an increasingly weaker scientific foundation. The main source of this level of influence seems to have been their political capital, particularly the movement's capability to mobilize the public in support of its position as well as a very active and vocal presence at the international scene. The wide arsenal of means they have at their disposal in this regard, from values and norms through persuasion and "shaming and blaming" to more coercive means have proved to be very effective in this regard. Because this is an issue area where almost no country has an economic interest, however, it seems reasonable to assume that environmental NGO influence is more easily achieved than on issues where countries face potentially high costs.

A fourth general conclusion is the significance of *allies*. A standard perception is that environmental NGOs are a weak but well-meaning underdog confronted by 'big industry' and powerful states as their main adversaries. This is by no means the case in the whaling issue. Apparently the two most powerful allies have been Greenpeace and the United States, and since 1980 there has not been any industry to fight at all. So some of the stereotypes associated with the role played by the environmental movement in international policy making should be scrutinized more critically.

Influence through alliances with state actors, however, has another interesting methodological aspect that makes it even more difficult for analysts to assess causality between NGO activities and outcomes. In the whaling case, our assessment could be "controlled" by asking the counterfactual question of whether the environmental NGOs were likely to have succeeded in getting the moratorium adopted in the absence of strong US support. Considering their mastery in framing the issue, their high level of access and participation both domestically and internationally combined with weak counterbalancing forces, they may well have succeeded.

Finally, it is interesting that the influence relations among non-state actors in the international whaling regime seem to be shifting again. The science-based approach to whaling management is gaining ground while support for a continued moratorium on commercial whaling has declined strongly. In contrast to earlier periods, the scientific basis for a change in policy has led to a high degree of scientific consensus. Even key (albeit "moderate") environmental NGOs such as WWF are now questioning the basis for maintaining the moratorium. Moreover, to an increasing extent, the whaling "community" is re-entering the process with full force after having been marginalized in the 1980s. This is a varied group of NGOs but with one thing in common, their pro-whaling stance. They operate independently of national governments, although some of the most influential ones are known to have close political and economic ties to key whaling states. No doubt the whaling NGOs have been important in legitimizing the pro-whaling stance, and they have been quite successful in undermining the previous "no-whaling" norm advocated by their NGO competitors. Currently, all three non-state actors are very active, but the pro-whaling forces have much more legitimacy than they have had in the past two or three decades. From the strong fluctuations in influence among these actors over time there is no guarantee that this development will continue. We should not be surprised if the anti-whaling forces strike back and once more change the curious game that takes place within the IWC.

Notes

Research for this chapter was conducted with support from the Research Council of Norway and the Political Science Department at the University of Oslo. An early version of this chapter was presented at the 44th ISA Convention, 26 February–1 March 2003, Portland, Oregon. This is a revised and updated version of an article published in *Global Environmental Politics* in 2003 (Skodvin and Andresen 2003) in which non-state influence in the IWC from 1970 to 1990 was analyzed. We gratefully acknowledge useful comments and suggestions to earlier drafts from Paul Wapner, Stacy D. VanDeveer, Elisabeth Corell, and Michele Betsill.

1. It should be noted that these shifts in influence were gradual and incremental and not as clear-cut as suggested here. This simplification, however, is useful for our analytical purposes.

2. This does not mean that non-state actors were the only important actors. Key nations, most notably the United States, had a significant impact, a point to which we return below.

3. The environmental movement and the animal rights movement differ in their arguments and positions on the whaling issue. We will not go into these distinc-

tions in any detail here. So for practical purposes we refer to this group as the environmental movement. Furthermore, while there is no doubt that a whaling industry once existed, it hardly makes sense to label current small-scale coastal whaling an "industry."

4. In 1959 Japan, Norway, and the Netherlands gave notice of their withdrawal from the Convention, but for different reasons. While Norway threatened to withdraw if the total quota was set above 15,000 units, the Netherlands pressed for an increase in the total quota. Schweder (2000: 83) notes: "Norway, still the largest whaling nation, and the Netherlands, eager to increase its catches, both left the IWC. As a consequence, the organization was at the brink of collapse."

5. A "blue whale unit" was equivalent to 1 blue whale, 2 fin whales, 2.5 humpback whales, or 6 sei whales (Schweder 2000).

6. For a discussion of the significance of alliances between powerful states and NGOs, see Gulbrandsen and Andresen, 2004.

7. For a detailed account of the scientific "battles" that were fought in the Scientific Committee during the 1950s, see Schweder (2000).

8. The United States has not adopted the same protectionist stance with regard to aboriginal whaling activities taking place within its own borders. In fact the United States had been a main pusher for a separation between aboriginal whaling and commercial whaling due to pressure from the aboriginals in Alaska.

9. These allegations were first put forth in a 1991 article in *Forbes Magazine* (Spencer, Bollwerk, and Morais 1991).

10. Pro-whaling countries, notably Japan, also used this strategy but not as successfully (DeSombre 2001).

11. Interestingly the previous IWC Secretary stated that information and opinions from this side were far more interesting and readable than the very predictable opinions and arguments of the anti-whaling NGOs (Andresen 2004).

12. This was part of the "Irish proposal" in 1997, which was an attempt to break the impasse between the two conflicting sides. For further elaboration, see Andresen (2001a).

13. Personal communication by Andresen with key decision-makers.

7

NGO Influence on International Policy on Forest Conservation and the Trade in Forest Products

David Humphreys

To what extent have NGOs exerted influence on international forest policy? This chapter explores this question by analyzing the activities of environmental NGOs and indigenous peoples' groups at international negotiations on forests and forest products. Two broad sets of values are common to the NGOs considered here: environmental values (with a particular emphasis on the ecological integrity of forests) and human rights values (with a particular emphasis on the rights of forest dwelling peoples). However, there are important differences between these NGOs. The NGO movement encompasses both system reformation and system transformation NGOs (Gale 1998a, b; Humphreys 1996b). Some NGOs adopt outsider tactics to target institutions, while others prefer an insider approach. The diversity and richness of NGOs campaigning on forests suggests that there is no coherent set of policy preferences across all NGOs.

The diversity among NGOs does not, however, rule out a set of shared concerns for protecting forests.[1] The majority of NGOs identify an urgent need to halt and reverse deforestation and biodiversity loss in all forested regions, and argue that reforestation should reproduce the original natural forest conditions as closely as possible. NGOs also contend that forest management should be ecologically sustainable and socially responsible over the long term. This includes the protection of old growth forests, sustainable yields of forest products, protection of endangered species and their habitats, elimination of clearcutting, protection of watershed and soil conservation functions, and the elimination of the use of chemicals. They argue that deforestation is often the result of incursions into the forest by powerful political and economic interests.

Halting deforestation therefore requires a redistribution of power relations from the global and national levels to the local level. Such a power shift is necessary to address many of the underlying causes of deforestation, including International Monetary Fund (IMF) structural adjustment programs, external debt and large-scale development projects.

NGOs also argue that the right to self-determination of local communities and indigenous peoples should be respected. The traditional and ancestral knowledge of forest peoples should be recognized as an intellectual property right, and such peoples should receive a share of any benefits that accrue from the commercial exploitation of such knowledge. Local communities and indigenous peoples should have the right to participate in genuinely democratic and decentralized decision-making processes at the national and international levels. The special role of women in forest conservation should be recognized. The rights of local communities and indigenous peoples to land should be recognized too, including the reform of inequitable distributions of land tenure (adapted from Donovan 1996: 94–95).

The international forests regime spans several organizations and instruments with a forest-related mandate (Glück et al. 1997; Tarasofsky 1999; Humphreys 2003). Apart from a brief consideration of the Convention on Biological Diversity (CBD), we do not consider hard legal instruments with a forest-related mandate but where forests are not the principal concern, such as the Kyoto Protocol and the Convention to Combat Desertification. Instead, our central focus is on multilateral negotiations in venues where forests or forest products have been the core concern. We examine forest negotiations at the United Nations Conference on Environment and Development (UNCED), the negotiations that took place under the auspices of the Commission on Sustainable Development (CSD) that produced "proposals for action" on forests in 1997 and 2000, and the consultation process that led to the creation of the United Nations Forum on Forests. We also consider two negotiation processes on forest products, namely negotiations on the international trade of tropical timber in the International Tropical Timber Organization and the discussions on forest products that took place under the auspices of the World Trade Organization in the late 1990s. The objective is to assess how far NGO activity has influenced the negotiations and the content of any textual outputs in these diverse institutional settings.

The positions adopted by NGOs have been ascertained principally from written statements circulated at the negotiations. Some NGO activists have been interviewed, both to check on the NGO positions as portrayed in the written statements and to seek a view from the activists on the influence they believe that NGOs have achieved. The outputs from the negotiation processes—the text of legal instruments and other formal UN documents—have been scrutinized to see if there is evidence of text that was first proposed by NGOs. In addition to original research, some use has been made of research carried out earlier (Humphreys 1996a, b, c). Primary and secondary source material was gathered and interviews conducted at the fourth session of the Intergovernmental Forum on Forests in 2000.[2]

The UNCED Forest Negotiations

By the late 1980s tropical deforestation was an important international issue, and one view in the NGO community was that a higher United Nations profile for the issue was needed. The European environmental NGOs network ECOROPA launched a petition and lobbying campaign in 1987. The objective of the campaign, which highlighted the cultural and ecological destruction of tropical forests resulting from forest-based industrial development, was for the United Nations General Assembly to convene an emergency special session on tropical deforestation. In September 1989 a petition of 3.3 million signatures was presented to the UN Secretary-General Javier Pérez de Cuéllar in New York (Meyer 1990). However, the agenda-setting campaign did not succeed in its objective of convening a General Assembly special session. The main reason for this was that the General Assembly was on the verge of formally moving to convene the UNCED in Rio de Janeiro. Three months after the petition was presented, the General Assembly passed resolution 44/228 announcing that the UNCED would be held in 1992. Among the issues that the resolution requested the UNCED to consider was forests.

In order to understand the influence that NGOs had on the UNCED forest negotiations, it is first necessary to understand the structure of the

negotiations. The forest negotiations saw considerable North–South polarization at the intergovernmental level reflected in debates over how to
frame the issue of deforestation. The developed North sought to construct deforestation as a "global" environmental issue. In as much as deforestation was an environmental issue, it was framed by the Group of
77 Developing Countries (G-77) as a national one. Claiming sovereignty
over their forest resources, the G-77 asserted their right to forest development and resisted what they perceived as Northern interference over
their forest resources. The G-77 used the forests negotiations to introduce issues that it had first raised in the 1970s, such as increased financial and technology flows from the North. Northern concessions, the
G-77 argued, were necessary in exchange for any commitments from
the South to tropical forest conservation. The North, unwilling to recognize the G-77's issue linkages, sought to keep the negotiations narrowly
focused on forest conservation and favored a forests convention, which
the South did not. The outcome was a non–legally binding statement of
forest principles.

This structure presented NGOs with both opportunities and con-
straints. On the one hand, the North–South polarization meant that government delegations were narrowly focused on their own interests and
on resolving conflicts with their "opponents." This constricted the political space within which NGOs could operate. On the other hand, the different framings of the forests issue between governments of the North
and South gave NGOs some room for maneuver. NGOs were able to
exploit the intergovernmental differences by lobbying different delegations on different issues, depending on the preferences of individual
delegations.

However, the opportunities for NGOs to take advantage of intergovernmental differences would have been enhanced had NGOs themselves
been more unified. There was no clear NGO view on whether there
should be a forests convention, with some NGOs inclining toward a convention that contained strict conservation targets and provisions on the
rights of indigenous peoples, while other NGOs opposed a forest convention (Humphreys 1996c: 100–101). Despite these divisions, NGOs
did realize some of their objectives. The negotiation of the forest principles was just one focus for forest NGOs; with most of the world's terres-

trial biodiversity found in tropical forests, forest NGOs also targeted the CBD negotiations. NGOs exerted some influence over both negotiation processes, and concerns with which NGOs have a long campaigning history, namely participation, indigenous knowledge, women's rights, and benefit sharing, appear in the final text of both documents (table 7.1).

What influence NGOs did achieve occurred in the early stages of the UNCED forest negotiations. NGOs successfully lobbied some delegations from the North to include text on these issues. However NGO proposals rarely survived intact; other delegations less sympathetic to NGO concerns would invariably make amendments to language. For example, NGOs successfully lobbied for mention of inequitable patterns of land tenure (i.e., the concentration of land ownership in the hands of economically and politically powerful interests). The inclusion of this issue in the forest principles is a clear sign of NGO influence as, in the absence of pressure from NGOs, government delegations had no incentive to provide international recognition of this issue. But this also meant that other delegations had an incentive to soften the original wording, lest recognition of inequitable patterns of land tenure should be construed as a criticism of government economic policy. This is what happened: delegates from the South caveated the wording proposed by NGOs so that the reworded clause in the forest principles refers only to the promotion "of those land tenure arrangements which serve as incentives for the sustainable management of forests" (United Nations 1992c, Principle 5).

As the negotiations progressed and intergovernmental positions hardened, delegates concentrated on resolving the core differences among them, further restricting the political space within which NGOs could operate. For example, prior to the final preparatory committee meeting NGOs submitted a proposed draft of the forest principles (NGO Statement at PrepCom 4 1992). However, at this stage the forest negotiations were mired in intergovernmental disagreement, and there was no space for NGOs to introduce original or innovative ideas. NGOs' lobbying efforts at this stage were negated by a North–South intergovernmental conflict on financial aid and the desirability of a post-UNCED forests convention. Tony Juniper of Friends of the Earth considers that at this late stage of the negotiations NGOs achieved no further influence on the text of the forest principles.[3]

Table 7.1
Text appearing in the UNCED forest principles and the Convention on Biological Diversity as a result of NGO lobbying

Issues	Non-legally binding statement of forest principles, 1992	Convention on Biological Diversity, 1992
Participation	"opportunities for the participation of interested parties, including local communities and indigenous people... nongovernmental organizations and individuals, forest dwellers, and women" [Principle 2(d)].	"allow for public participation in such procedures," i.e., impact assessment and minimizing adverse impacts [Article 14.1(a)].
Indigenous knowledge	"Appropriate indigenous capacity and local knowledge regarding the conservation and sustainable development of forests" should be recognized, respected, recorded, and developed [Principle 14(d)].	"respect, preserve and maintain knowledge, innovations and practices of indigenous and local communities embodying traditional lifestyles" [Article 8(j)].
Role of women	"The full participation of women in all aspects of the management, conservation and sustainable development of forests should be actively promoted" [Principle 5(b)].	"affirming the need for the full participation of women at all levels of policy-making and implementation of biological diversity conservation" [Preamble].
Benefit sharing	"Benefits arising from the utilization of indigenous knowledge should therefore be equitably shared with such people," i.e., people in the local communities concerned [Principle 14(d)].	"the desirability of sharing equitable benefits arising from the use of traditional knowledge, innovations and practices relevant to the conservation of biological diversity" [Preamble. Also mentioned in Article 1 and Article 8(j)].

Sources: United Nations 1992c (column 2) and 1992b (column 3).

Overall, the UNCED forests and biodiversity negotiations were partial successes for forest NGOs. The negotiations represent the start of a trend that has continued into the post-UNCED era where NGOs have succeeded in placing issues on the agenda, and in getting some of their concerns inserted into negotiated texts. However, the language that NGOs initially propose is often later weakened by delegates so that the substance of the final text is diluted, with substantive commitments avoided. The influence of forest NGOs in this institutional context falls somewhere between moderate and high according to the criteria introduced in chapter 2.

The CSD Forest Process, 1995 to 2000

By the conclusion of the UNCED negotiations there was little goodwill or confidence at the intergovernmental level between North and South. NGOs could do little to affect this situation, which could only be resolved between governments. The stalemate was broken in 1994 when an intergovernmental working group co-sponsored by the Canadian and Malaysian governments established an international agenda for forestry that was adopted in modified form by the CSD. This agenda led to the creation of two temporary CSD subgroups: the Intergovernmental Panel on Forests (IPF) from 1995 to 1997 and its successor, the Intergovernmental Forum on Forests (IFF) from 1997 to 2000.

Intergovernmental Panel on Forests

Not surprisingly, the agenda for the IPF agreed in 1995 strongly reflects the unfinished business of the UNCED, and includes issues such as financial and technology transfers and the relationship between trade and the environment. However, NGO lobbying also influenced the agenda in at least two instances.

The first is an issue on which forest NGOs have a long campaigning history, namely the contribution that "traditional forest-related knowledge" (TFRK) can make to sustainable forest management. Having successfully lobbied for inclusion of language on indigenous knowledge in the UNCED forest principles and the Convention on Biological Diversity, NGOs lobbied for inclusion of the issue on the IPF agenda (United

Nations 1995: 3). No single NGO statement or NGO action can be identified as the source of NGO influence on this issue. Its inclusion was the result of consistent campaigning by numerous forest NGOs over several years. Still it needs to be noted that some governments from the South have an incentive to recognize the concept of TFRK. Rosendal (2001b) argues that although TFRK has been promoted primarily by NGOs, as an issue, it ranks highly on the G-77 agenda since it asserts the importance of protecting the intellectual property rights of tropical forest countries over their genetic inheritance. NGOs achieved influence on this issue in part because their concerns resonated with the interests of key states. So, by the time the IPF was created, NGOs were pushing against an open door (Humphreys 2006).

The second example of NGO influence, which concerns the mandate of the IPF in clarifying international institutional arrangements, can be traced to a single NGO statement. At the 1995 CSD meeting that established the IPF, a statement was submitted by the Global Forest Policy Project (GFPP) (1995: 1), a project of the National Wildlife Federation, Sierra Club and Friends of the Earth–US. "Global Forest Policy Project respectfully submits the following recommendations for consideration by members": the IPF would "[c]ommission a thorough review and assessment, *by an independent high-level body*, of existing international instruments and institutions concerned with forests and related matters" [emphasis in original]. The review would "[i]dentify overlaps and redundancies..." and "[i]dentify gaps where existing instruments or institutions appear insufficient to address important forest policy issues and problems...." The final text subsequently issued at the end of the CSD meeting stated that the IPF would "[d]evelop a clearer view of the work being carried out by international organizations and multilateral institutions...in order to identify any gaps, and areas requiring enhancement, as well as any areas of duplication" (United Nations 1995: 4–5).

The influence of the GFPP with respect to the language on overlaps/ duplication and gaps is clear. However, the full import of the GFPP did not carry through into the CSD text, as the emphasis on an independent high-level body was lost. Furthermore the GFPP recommendations also contained a list of institutions and instruments the IPF could consider, including International Labor Organization (ILO) Convention 169, which recognizes the rights of indigenous and tribal peoples to determine their own land use policy. Nevertheless, the CSD did not adopt a reference to this ILO convention. Persistent NGO lobbying for the CSD to recognize and work within the provisions of ILO Convention 169 met with failure, in large part because there is a conceptual tension between the rights of indigenous peoples to determine their own land use policy and the sovereign rights of states to determine their natural resource policy. In intergovernmental negotiations the latter invariably prevails.

In order to assess the NGO influence elsewhere in the CSD process on forests, a brief understanding of how the IPF and its successor, the IFF, functioned is necessary. Both the IPF and the IFF met four times each, for two-week periods. Despite the limited time spent in formal negotiations, both the IPF and the IFF produced lengthy proposals for action that, as the name suggests, take the form of suggestions and recommendedations, principally for governments but also for international institutions (Humphreys 2001). These outputs were possible due to the work that took place between formal sessions. Specific thematic issues were addressed in various intersessional initiatives, which were sponsored by one government or more. Intersessional meetings tended to be informal and were open to NGOs and indigenous peoples' groups, an arrangement that was welcomed by the World Rainforest Movement as it enabled NGOs to broaden the international forest policy debate (Griffiths 2001).

Between the third and fourth sessions of the IPF, NGOs, in cooperation with the governments of Colombia and Denmark, organized an "International Meeting of Indigenous and Other Forest-Dependent Peoples on the Management, Conservation and Sustainable Development of all Types of Forests" in Leticia, Colombia. This generated the Leticia Declaration, which re-emphasized earlier demands made by indigenous peoples, including "[t]hat the rights, welfare, viewpoints and interests of Indigenous Peoples and other forest-dependent peoples should be central to all decision-making about forests at local, national, regional and international levels" (United Nations 1997c: 10). The Leticia meeting also produced some draft proposals for action for consideration by the IPF at its fourth session. A comparison between the Leticia Declaration and the text adopted by the IPF at its fourth session reveals that the former had an impact on the latter.

Consideration cannot be given here to all issues and all text proposed by the Leticia Declaration. However, the case of TFRK is again illustrative. The Leticia Declaration advocated that use of TFRK "should not be made without the prior informed consent of the Peoples concerned" (United Nations 1997c: 13). Subsequently the final report of the IPF states that:

Governments and others who wish to use TFRK should acknowledge, however, that it cannot be taken from people, especially indigenous people, forest owners, forest dwellers and local communities, without their prior informed consent. (United Nations 1997b: para. 36)

The principle of prior informed consent advocated by the Leticia Declaration thus influenced the final IPF report. However, there was some weakening of the Leticia proposal. First, the Leticia emphasis on "Peoples" (uppercase, plural) was changed in the IPF report to "people" (lowercase, singular), thus denoting a lower status for indigenous peoples. Second, the emphasis in the Leticia Declaration on indigenous peoples was broadened to include the knowledge of other actors. Indeed the IPF stressed that "TFRK should be broadly defined to include not only knowledge of forest resources but also knowledge of other issues that are considered relevant by countries based on their individual circumstances" (United Nations 1997b: para. 32). Such an emphasis thus admits agencies that have been criticized for promoting forest loss, such as industrial timber companies. But NGOs had promoted the concept of TFRK because they wished the often-excluded life protective knowledge of forest peoples to be admitted to decision making, and the IPF adoption of a broad interpretation to include the knowledge of actors from outside the forests thus weakened the concept.

At the IPF negotiations the conclusion is that, once again, NGOs exerted a moderate to high level of influence according to the criteria introduced in chapter 2.

Intergovernmental Forum on Forests
In 1997 the IPF was replaced by the IFF. Like the IPF, the IFF was a CSD subgroup; indeed to all intents and purposes it was the IPF with a revised agenda. NGOs continued to attend the formal negotiation sessions and to play an active role in the intersessionals. One intersessional initiative

in which NGOs played a leading role was on the underlying causes of deforestation, which was also an agenda item for the IPF. The task manager for this issue was the United Nations Environment Programme (UNEP), although it was the NGO community that seized the initiative. At the first meeting of the IFF, NGOs announced their willingness to contribute to research on the international and national causes of deforestation. Subsequently seven regional workshops and one workshop of indigenous peoples' organizations were held. The results fed into a global workshop in 1999 hosted by the government of Costa Rica and co-organized by NGOs and UNEP. The final report of the initiative, coordinated by the Biodiversity Action Network, was presented to the third session of the IFF (Verolme and Moussa 1999).

A comparison between the text proposed by the NGOs and the report of the IFF reveals findings that are similar to those on the impact of the Leticia Declaration at the IPF. Some text proposed by the NGOs was adopted, while other proposals were either weakened or not adopted at all. For example, some of the causes of deforestation reported by the NGO initiative appear in the IFF's report, although the text was substantially modified. The report from the NGO workshop stated, *"The non-recognition of the territorial rights of indigenous and other traditional peoples*, and the resulting invasion of these territories by external actors was often highlighted as an underlying cause" (Verolme and Moussa 1999: 3; emphasis in original). The IFF's report notes the role in deforestation of "inadequate recognition of the rights and needs of forest-dependent indigenous and local communities within national laws and jurisdiction" (United Nations 2000: para. 58). In another example, the NGO report emphasized *"the lack of empowerment and participation of local communities in decisions over forest management"* (Verolme and Mousa 1999: 5; emphasis in original), while the IFF language simply refers to a "lack of participation" (United Nations 2000: para. 58).

Much language in the NGO report does not appear at all in the IFF's report. The NGOs emphasized the following as underlying causes of deforestation: *"Government led colonization processes into the forests*, stemming from *inequitable land tenure patterns"* (Verolme and Mousa 1999: 3; emphases in original). The IFF makes no criticism of government policies (not surprisingly, given the sensitivity of many G-77 governments on sovereignty over their resources) but does note the role of "lack of secure land tenure patterns" in deforestation (United Nations 2000: para. 58). The role of inequitable land tenure arrangements as a driving force of deforestation has been a consistent theme in the forest NGO literature.

NGOs were also unsuccessful in their proposal for the IFF to recognize as an underlying cause "*the privatization of forests for the benefit of large-scale private or corporate landowners*" (Verolme and Mousa 1999: 4; emphasis in original). Indeed the IFF took the opposite view on the private sector and emphasized instead the role of private sector finance in forest projects. "The concept of an international investment promotion entity to mobilise private sector investment in SFM [viz. sustainable forest management] deserves further consideration" (United Nations 2000). The IFF's report is also replete with mention of "public-private partnerships," a concept that has acquired a prominent place in contemporary neoliberal discourse, and that captures the idea that the state should mobilize private money to replace declining public expenditure on public goods and public services. NGO textual proposals tend to be blocked where they run directly counter to neoliberal discourse. Government delegates screen NGO textual proposals for language that challenges the core interests of the global economy, so such language is selected against in the negotiation process.

The IFF report mentions the concept of multistakeholder dialogue (United Nations 2000: Annex, para. 5). NGOs at the IFF had lobbied for this concept, although its inclusion was not solely the result of NGO advocacy in forests negotiations. The multistakeholder dialogue concept was first adopted at the second United Nations Conference on Human Settlements (Habitat II) in 1996 and was later adopted by the CSD.[4] The adoption of the concept by the IFF was the cumulative result of lobbying by many NGOs, including, but not solely, forest NGOs. Multistakeholder dialogue is increasingly being adopted outside the UN system. For example, the Ministerial Conference on the Protection of Forests in Europe held in Vienna in 2003 included a multistakeholder dialogue segment.

NGOs also exerted influence on other issues in the IPF/IFF process. Tom Griffiths (2001: 4) of the World Rainforest Movement argues that of the approximately 300 IPF/IFF proposals for action produced, "78 address indirectly or directly indigenous proposals on land tenure, participation, cross-sectoral policies, community forest management and traditional forest-related knowledge." However, there were areas where NGOs had less success. NGO proposals that were not incorporated in the proposals for action relate to demands for autonomy and self-determination and the mainstreaming of international law on indigenous peoples' rights into international forest policy (Griffiths 2001: 4). This is another example of NGOs exerting a moderate to high level of influence.

Bill Mankin of GFPP attributes the success of NGOs in the IPF/IFF process to the following tactics:

Where NGO representatives have had particularly close and cordial relationships with one or more delegates/delegations, where they [NGOs] have been closely dogging the negotiation of very specific language step-by-step and hour-by-hour, where the NGO reps have been well-respected by the delegates, and where the NGO reps have been pretty good wordsmiths, they've often had success in getting language added, deleted, or changed.[5]

By the end of the IFF process, NGOs held the view that there should be no further political negotiations on additional proposals for action; governments instead should concentrate on implementing the existing proposals for action. While the language in the agreed proposals was weaker than what the NGOs would have liked, most NGO campaigners felt that they would gain greater influence by monitoring and contesting the implementation of these proposals within countries and through national reporting at the UN than could be gained from further multilateral negotiations.[6]

United Nations Forum on Forests

After the IFF was dissolved in 2000, members of the United Nations decided to create a new organization, the United Nations Forum on Forests (UNFF). Unlike the IPF and IFF, the UNFF does not report to the CSD but reports directly to the United Nations Economic and Social Council. As they have with previous international initiatives on forests, NGOs sought to influence the agenda of the UNFF from the very start. The UNFF's agenda was largely shaped by an initiative hosted by Germany called the 8-Country Initiative.[7] This initiative, which was held in consultation with the UNFF secretariat, concluded with a workshop in Bonn where NGO activists stressed that "existing national–international reporting mechanisms are either ineffective or inappropriate to assess the implementation of the IPF/IFF Proposals for Action" and that the "UNFF must do something original to foster and promote implementation" (World Rainforest Movement 2000). One activist, Marcus Colchester, noted that the workshop discarded "key civil society recommendations relating to the need to apply a bottom up approach to the UNFF focus on implementation based on monitoring and reporting involving NGOs, IPOs [indigenous peoples' organizations] and civil society" (World Rainforest Movement 2000).

It is thus no surprise that when the UNFF process began in 2001, it tended toward the politics of the lowest common denominator. Australia had spoken in favor of compulsory national reporting, and proposed that countries "will" report on implementation. The United States, supported by some developing countries, argued for voluntary national reporting. The United States also argued against collectively agreed implementation targets, and insisted that countries alone should set individual targets and timetables (IISD 2001). The UNFF states that targets would be "set by individual countries within the framework of national forest processes, as appropriate" (United Nations 2001: 18). Overall, NGOs had a low level of influence on the UNFF negotiations; they appear to have had minimal impact on the negotiation process or outcome. With countries slow to submit national reports, some NGOs have considered abandoning the UNFF (World Rainforest Movement 2002). As NGO activist Bill Mankin has argued, the record of the IPF and IFF on implementation is an "indictment" of the "entire post–UNCED forest debate."[8]

International Trade in Forest Products

In addition to negotiations on forests, various international negotiations have been held on forest products. Here we consider negotiations on the international trade of tropical timber within the context of the International Tropical Timber Organization and the proposal for a forest products agreement under the auspices of the World Trade Organization.

International Tropical Timber Organization

The International Tropical Timber Organization (ITTO) was created by the International Tropical Timber Agreement (ITTA) of 1983. NGOs exerted a high level of influence on the negotiations that produced this agreement. They reframed the issue of tropical timber and successfully lobbied to have their substantive and procedural concerns reflected in the final treaty text. NGOs challenged the dominant framing of tropical timber as a resource and development issue at an intergovernmental meeting held before the main negotiating conference when the IUCN,[9] supported by other NGOs, argued that the agreement should recognize the importance of forest conservation and its relationship to tropical forest development. States subsequently agreed to this and inserted into the agreement a clause that Parties should aim at the "sustainable utilization of and conservation of tropical forests and their genetic resources, and at maintaining the ecological balance in the regions concerned" (United Nations 1983, Article 1.h). The 1983 agreement was superseded by the International Tropical Timber Agreement of 1994, which itself was replaced by the International Tropical Timber Agreement of 2006. The conservation emphasis of the 1983 agreement was retained in both of these successor agreements. The three agreements are the only international commodity agreements to contain a conservation clause. NGO lobbying at the original ITTA negotiations also led to some important participation rights at the ITTO. At its first meeting, the ITTO agreed that any NGO attending the semiannual meetings should be granted observer status, as should national timber trade organizations.

Four years after the ITTO started operating, Friends of the Earth–UK drafted a proposal for timber labeling whereby tropical timber from sustainable sources would receive an ITTO label certifying sustainability. The NGO proposal was formally tabled by the UK delegation (ITTO 1989a). However, the proposal was blocked after objections from Indonesia and Malaysia, the latter stating that the proposal was "a veiled attempt... to encourage the current campaign of boycott" against the international tropical timber trade (ITTO 1989b: 6). Despite successfully influencing the UK delegation, Friends of the Earth was unsuccessful in getting the ITTO as a whole to adopt its labelling proposal because the proposal lacked support from key delegations in the producers' caucus. In this instance NGOs had a moderate level of influence, according to the criteria introduced in chapter 2. They shaped the negotiation process within the ITTO by influencing the position of a key state (the United Kingdom) but ultimately failed to have their position adopted by the organization. The failure of the ITTO to adopt a timber labeling scheme catalyzed the efforts of the World Wide Fund for Nature (WWF), along with other NGOs and timber traders, to create the Forest Stewardship Council, an independent voluntary timber certification scheme. NGOs have therefore had more success in modifying the business agenda on certification and labeling than they have had on the intergovernmental agenda.

One significant ITTO landmark was NGO research by the International Institute for Environment and Development (IIED) on sustainable forest management. In 1988 the IIED reported that only one-eighth of one percent of the world's tropical forests (in Queensland, Australia) provided timber managed from sustainable sources (Poore 1989). These findings led WWF to call for the ITTO to agree that by 1995 the international trade of all timbers—tropical and nontropical—should come from sustainable sources. An important route of influence for the WWF was its status with key government delegations. As a trusted insider NGO, WWF advisers were appointed to the delegations of three countries: United Kingdom, Denmark, and Malaysia. This gave the WWF an important route of influence, with these advisers being privy to intergovernmental discussions taking place behind closed doors from which NGO observers were excluded. The WWF's proposal for a 1995 target year for all timbers was supported by other NGOs but was not adopted by the ITTO, which in 1990 opted instead for the target year of 2000 applied only to tropical timbers. NGOs thus had some success on this issue, although WWF's recommended target date was adjusted by five years and the emphasis on nontropical timbers was lost. On this issue NGOs achieved a moderate level of influence.

Once the 2000 target date had been adopted by tropical timber producing states NGOs lobbied for it to be adopted by other timber producing states that were ITTO members. NGOs pressed for this during the negotiations for the second International Tropical Timber Agreement that took place between 1992 and 1994. The NGOs were supported by the tropical timber producers, although it was clear that the proposal would be adopted only if nontropical governments could be influenced. The US government was one of the first to support the NGO position and made the decision after considerable pressure from American NGOs, in particular, the WWF-US (although the National Wildlife Federation, the Sierra Club, and Friends of the Earth–US also played a role; see Humphreys 1996a). Toward the end of the negotiations other consumer governments conceded on this point. The consumers' commitment to the 2000 target is noted in the preamble to the International Tropical Timber Agreement of 1994 (United Nations 1994b).

NGOs failed nevertheless to achieve their main objective in these negotiations, namely an expansion in the scope of the new agreement to include all timbers. Again, NGOs were supported in this aim by the producer countries. But they were opposed by the consumer countries, which wished to retain the tropical-timber-only scope. Shortly after the start of the negotiations, WWF announced that it was withdrawing its advisers from all government delegations in order to apply pressure on delegations to expand the scope of the new agreement, to signal discontent at the ITTO's poor conservation record, and to highlight the absence of NGOs from other delegations. The consumers' view ultimately prevailed after consumer delegations indicated that their willingness to continue funding projects in tropical timber countries depended on the scope of the agreement remaining unchanged. With the negotiations polarized between producers and consumers, the political space within which NGOs could maneuver was limited. This led to some frustration among the NGOs, with one campaigner, Bill Mankin of the Global Forest Policy Project, commenting that "NGOs had worked hard to convince members that the negotiations offered an opportunity to chart a new course. Yet it was hard to tell whether NGOs' textual proposals had been seriously considered or not" (Mankin, cited in Poore 2003: 125).

By the end of the negotiations it was clear that NGOs had successfully exerted influence on issues beyond the 2000 target. NGOs lobbied for the agreement to promote reforestation, and to recognize the rights of local communities. The 1994 agreement does not mention the word "rights" with respect to local communities, but it does go some way to meeting the concerns of the NGOs. Article 1(j) states that Parties should encourage "reforestation and forest management activities as well as rehabilitation of degraded forest land, with due regard for the interests of local communities dependent on forest resources" (United Nations 1994).

Although NGOs have achieved some influence at the ITTO, most NGOs have been disappointed that they have not achieved more. In confronting a powerful coalition of timber trade federations and timber producing states, NGOs have been unable to ensure that environmental considerations prevail over timber trade interests. In fact frustration that the ITTO was not expandable into an International Timber Trade Organization with responsibility for all timbers has caused many international NGOs to cease attending meetings of the ITTO after the conclusion of negotiations for the International Tropical Timber Agreement, 1994 (Gale 1998b).[10]

World Trade Organization

Toward the end of the 1990s the World Trade Organization (WTO) initiated work on a Forest Products Agreement. The US administration was one of the main supporters of the proposed agreement, whose aim was to eliminate tariffs on all wood products as part of America's trade liberalization agenda (Madeley 2000). When NGOs learned that a Forest Products Agreement could be signed as early as the 1999 WTO meeting in Seattle, they initiated a campaign against the agreement. A statement signed by 140 NGOs worldwide was circulated to the WTO and the CSD stating:

We condemn the proposal because, if implemented, we fear it will lead to increased logging and consumption of ecologically and socially valuable forests around the globe. We also condemn the proposal because of the undemocratic and ecologically irresponsible manner in which it is being developed. (FERN 1999)[11]

In response to these concerns, the US government announced in July 1999 that it would conduct an analysis of the economic and environmental impacts of the proposed agreement (Federal Register Notice 1999). However, the United States did not announce that it would halt negotiations, which at this stage were taking place behind closed doors.

The Forest Products Agreement was not signed at the Seattle meeting of the WTO, nor at the time of writing (March 2007) has the issue re-emerged at subsequent WTO meetings. Deforestation was one issue among many that led to the massive street protests against the WTO at Seattle and that resulted in several negotiation sessions being abandoned.

Here we can pose a counterfactual question: What would have happened if NGOs had not opposed the proposed Forest Products Agreement? We have seen that there was NGO lobbying and agreement was not concluded. However, it cannot be concluded that the Forest Products Agreement was dropped *because of* NGO pressure, either the campaign by forest NGOs prior to Seattle or the demonstrations by NGOs and the nascent global justice movement on the streets of Seattle. The lack of transparency with which the WTO operates means that establishing the causes for the apparent abandonment of the Forest Products Agreement cannot be done with certitude. NGO campaigning certainly led to some reconsideration of the US position, at least in so far as the position of the US government can be judged from its public policy statements, although there was no indication prior to Seattle that the US or the WTO was planning on abandoning the agreement.

Reasons other than NGO campaigning may explain why the agreement has not been concluded in the WTO. First, the Forest Products Agreement was a relatively low trade concern for core governments, and since Seattle the priority has shifted toward completing the General Agreement on Trade in Services (GATS). Second, the US government appears to have shifted its tactics with respect to the negotiation of international law on forest products. A forest products agreement is one of the priorities of the US delegation in negotiations for the proposed Free Trade Area of the Americas, which one US NGO, the American Lands Alliance, has condemned as a "free logging agreement" (American Lands Alliance 2000). Third, since Seattle, the WTO has become increasingly preoccupied with the question of agricultural subsidies, which has become a major concern for developing countries.

This analysis suggests that NGOs had a moderate level of influence on the WTO negotiations on a forest products agreement. NGOs likely had some effect on the negotiation process by shaping the US position. However, it does not appear that NGOs can be credited with blocking the final agreement.

Conclusions: Redefining the Issue

A number of conclusions emerge from this study. First, NGOs have a range of resources at their disposal. Particularly important here is their moral status as concerned independent organizations with autonomy and independence from other actors. This gives NGO research and arguments an authoritative status.

Second, the earlier NGOs become involved in an international negotiation process, the more likely they are to be able to influence that process. For example, NGOs influenced the contents of the 1983 International Tropical Timber Agreement and the agenda of the Intergovernmental Panel on Forests because they were involved in the negotiations at a sufficiently early stage. However, the more advanced a negotiation process is, the more difficult it is for NGOs to achieve textual changes, especially where key conflicts remain between delegations that require resolution. Possibly one reason why NGOs are showing disillusionment with the UNFF is that they were unable to influence significantly its agenda from the outset, and they now see limited opportunity to shift the terms of debate.

Third, the forests regime overlaps with other international environmental processes and fora, and concepts introduced into these other arenas can influence forest negotiations (and, more tentatively, vice versa). The concept of multistakeholder dialogue was adopted at the IFF, but it did not originate from there. While the concept of traditional knowledge appears in the forest principles, it was the stronger formulation of this concept in the Convention on Biological Diversity that was invoked as a precedent when the concept was discussed at the IPF and IFF. Where such institutional overlap exists, the contents of one environmental regime may thus affect those of others (Andersen 2002). In the case of forests, rules and principles adopted in one legal instrument may subsequently be adopted in others, resulting in a "spillover effect," with some rules and principles finding expression in several legal codes (Humphreys 2003). So, in order fully to assess the role of NGOs on the international forests regime, a thorough understanding of the role that NGOs have in all forest-related negotiating arenas would be necessary.

Fourth, in the short term, NGOs are more likely to influence the textual outputs of international forest negotiations if they frame their recommendations in language that is congruent with mainstream neoliberal discourse and that does not oppose the powerful political and economic interests that have found representation in state delegations. NGO textual proposals are screened by government delegates, and language that is too critical of governments or private interests allied to governments will be blocked. This is not to suggest that NGOs should blunt their critiques. On the contrary, the normative and critical force of NGO statements and arguments comes from their independence and resistance to compromises. However, in the short term the more radical textual proposals of NGOs will at best be modified, and at worst they will be completely filtered out of the negotiation process.

The emphasis on the short term in the previous paragraph is deliberate. The longer term influence of NGOs' more radical proposals is considerably harder to judge. In the short term tracking and assessing influence is relatively easy; one can compare NGOs' textual proposals with the final negotiated text, and look for evidence that the former has impacted upon the latter, as parts of this chapter have sought to do. However, formally tracking NGO influence over the long term is far more problematic, since establishing cause–effect relationships becomes more difficult when the time span between cause and effect increases and when more variables are introduced (e.g., the same proposals being made several times by different NGOs in many different fora). Moreover, influencing text is no guarantee that action on the ground will subsequently be taken. Hence forest NGOs are increasingly emphasizing that those textual outputs agreed to date should now be implemented, with governments reporting on national level implementation to the UNFF.

Another factor that complicates the assessment of NGO influence relates to the co-option of concepts. Some NGO concepts have been accepted by government delegates because they can be manipulated and used by core political interests for their own ends. We have seen that the concept of traditional forest-related knowledge is one such example. Furthermore it is not entirely clear that concepts such as participation and multistakeholder dialogue have been adopted because NGOs have successfully engineered a value shift in favor of more inclusive and democratic governance, or that the concepts have, at least in part, been adopted by developed world governments because they fit with the emphasis in neoliberal discourse on the declining role of the state in the economy along with the concomitant emphasis on non-state provision of public goods and services that enhance the roles for private business, the voluntary sector, charities, and so on. NGOs will use concepts such as multistakeholder governance and participation to pry open political processes for civil society, business and private sector actors will use the same concepts to gain a louder decision-making voice for themselves, while governments will be content to see other actors stepping forward to assume functions that were previously the domain of the state. The question then becomes who has the most power in such "open" and "transparent" dialogues.

With these reservations in mind, some of the main NGO achievements on the forests issue are summarized in table 7.2. The methodology of this volume leads to the conclusion that the level of influence that NGOs have achieved in negotiations on forests and forest products is high. NGOs have been actively engaged in forests negotiations for more than two decades. As the negotiations have moved among various institutional venues, NGO activities have produced observable effects on negotiation processes and outcomes. NGOs have succeeded in shaping how forest-related issues are framed, influencing the positions of key states and placing issues on the negotiating agenda. In a number of instances we find that NGO goals are reflected in the text of final agreements on both procedural and substantive issues.

Nevertheless, some qualifications need to be made to this conclusion. First, given the number of international processes there have been on forests over the last twenty years, the evidence presented in this study and summarized in table 7.2 can only be partial. But while the evidence is not exhaustive, it is indicative of broader trends. We have seen that NGOs have successfully placed new forest-related issues on the international agenda and have inserted language into negotiated text on many of their key concerns: these include participation, the role of women, traditional forest-related knowledge, benefit-sharing, and land tenure security. NGOs have certainly met with some failures, and often their "successful" proposals have been heavily caveated and weakened during negotiations. Furthermore there has been a tendency by some delegations to agree to some concepts as buzzwords, changing the NGO phrasing to yield language that is softer and more ambiguous and thus depriving the concepts themselves of any substantive meaning. Overall, however, NGOs have won for themselves a very respected place at international forest negotiations, and their contributions are valued and taken seriously.

Arguably the most important contribution that NGOs have made is the reframing of the issue of forest conservation from a purely economic issue to an ecological and human rights one. The theory of social construction is helpful here. Angela Liberatore (1995) has identified three "cognitive frames" for deforestation. The interpretations yielded by these frames may vary according to geographical location, culture, social conditions, the interests at stake, and the values of the actor to whom the issue is salient. An economic growth cognitive frame will view deforestation "in terms of cost–benefit calculations, trade conditions, and contributions to the gross national product." An ecological cognitive frame sees deforestation in terms of its disruption of environmental functions, while a human rights cognitive frame will perceive "deforestation as a danger for and crime against indigenous populations" (Liberatore 1995: 68).

All three cognitive frames can be discerned in international forest negotiations. In the early negotiations for the International Tropical Timber Agreement of 1983 forest use was constructed by governments as a resource issue. This fits with an economic growth cognitive frame. Forest NGOs have been active in lobbying for a reframing of the issue in line with ecological and human rights framings. Despite reversals, NGOs have had considerable success in promoting the ecological and human

Table 7.2
NGO influence on international negotiations over forests and forest products

	Influence indicator	Evidence		NGO influence? (yes/no)
		Behavior of other actors…	…as caused by NGO communication	
Influence on the process	Issue framing	Government delegations no longer view forests principally as an economic resource. They have increasingly recognized the ecological and human rights dimensions of forests use over the last twenty years.	NGOs have persistently lobbied for forest conservation and for the rights of forest peoples. They have transmitted their concerns on these issues to a range of international fora dealing with forests and forest products.	Yes
	Positions of key actors			
	• United Kingdom	*International Tropical Timber Organization:* The UK delegation tabled a proposal for the labeling of tropical timber from sustainable sources.	The proposal had been drafted by Friends of the Earth–UK.	Yes
	• Producer countries	*International Tropical Timber Organization:* Producer countries agree to adopt the target date of 2000 by which time tropical forests contributing to the international tropical timber trade should come entirely from sustainable sources.	Led by the World Wide Fund for Nature, NGOs lobby for all countries to adopt a target date of 1995 by which time tropical and nontropical forests contributing to the international timber trade should come entirely from sustainable sources.	Yes

• United States	*Negotiation of International Tropical Timber Agreement, 1994:* US delegation is the first to argue for adoption by consumer countries of the 2000 target. Other consumer delegations subsequently agree.	All NGOs taking part in the negotiations advocate consumer adoption of the 2000 target. US NGOs had separately lobbied the US government on this subject.	Yes
	World Trade Organization: US government agrees to conduct an analysis of the economic and environmental impacts of a WTO Forest Products Agreement.	NGOs vigorously oppose the proposed Forest Products Agreement.	Yes
Agenda setting	*Intergovernmental Panel on Forests:* IPF agenda includes assessing the work carried out by international organizations in order to identify gaps, areas requiring enhancement and areas of duplication.	A proposal by the Global Forest Policy Project to the 1995 CSD meeting recommends that the IPF address these issues.	Yes
Influence on negotiating outcome	Final agreement/ procedural issues	*UNCED:* The forest principles recognize that local communities, indigenous peoples and women should have the right to participate in forest policy. These principles also appear in the IPF/IFF proposals for action.	NGOs have a long campaigning history on these issues and lobbied actively for them during the negotiations.	Yes

Table 7.2
(continued)

Influence indicator	Evidence		NGO influence? (yes/no)
	Behavior of other actors…	…as caused by NGO communication	
	Intergovernmental Forum on Forests: Deforestation is noted to take "inadequate recognition of the rights and needs of forest-dependent indigenous and local communities within national laws and jurisdiction."	The NGO report on underlying causes emphasizes that the "non-recognition of the territorial rights of indigenous and other traditional peoples, and the resulting invasion of these territories by external actors was often highlighted as an underlying cause."	Yes
	International Tropical Timber Agreement, 1994: The agreement recognizes "due regard for the interests of local communities dependent on forest resources."	NGOs had lobbied for the agreement to recognize the "rights" of local communities.	Yes
Final agreement/ substantive issues	*Intergovernmental Panel on Forests:* TFRK should not be taken from forest people without their "prior informed consent."	The Leticia Declaration of NGOs advocates that use of traditional forest-related knowledge "should not be made without the prior informed consent of the Peoples concerned."	Yes

Intergovernmental Panel on Forests and *Intergovernmental Forum on Forests:* Analysis by the World Rainforest Movement reveals that 78 of the IPF/IFF proposals for action contain language on land tenure, participation, cross-sectoral policies, community forest management, and traditional forest-related knowledge.	NGOs had a long campaigning history on these issues before the establishment of the IPF. They lobbied persistently on these issues at the IPF (1995–1997) and IFF (1997–2000).	Yes
International Tropical Timber Agreement, 1983: Parties should aim at the "sustainable utilization and conservation of tropical forests and their genetic resources." A similar clause appears in the 1994 and 2006 agreements.	The IUCN, supported by other NGOs, stated that that the agreement should recognize the long-term relationship between forest conservation and forest development.	Yes
Level of influence		High

rights dimensions of forest use in a range of fora and institutions. Issue definition not only reflects the patterns of values and interests at play in negotiations. It is also crucial in determining the way in which an issue is subsequently handled. Over the long term NGOs have played—and will continue to play—an important role in helping to shift patterns of values and interests.

Notes

This chapter is an amended and revised version of a paper published earlier in *Global Environmental Politics*, vol. 4, no. 2, 2004, pp. 51–74.

1. The existence of a set of common demands does not, of course, prevent disagreement and divisions between NGOs, nor rule out the need for dialogue and discussion on campaigning priorities, strategy, and tactics. This brief summary does not capture all NGO views, although it is indicative of their demands in international negotiations on forests.

2. I am grateful to UNED–UK who kindly provided accreditation for this meeting, and to the Faculty of Social Sciences at The Open University who funded the visit.

3. Tony Juniper, Rainforest Campaigner, Friends of the Earth–UK, London, personal communication (interview), London, 12 August 1992.

4. I am indebted to Felix Dodds of the Stakeholder Forum for explaining the origins of the multistakeholder dialogue concept; personal communication (interview), Stockholm, 29 August 2003. See Hemmati (2003).

5. Bill Mankin, Global Forest Policy Project, personal communication (email), 1 August 2003.

6. Marcus Colchester, Forest Peoples Programme, personal communication (interview at fourth session of the IFF, New York), 8 February 2000. Bill Mankin, Global Forest Policy Project, personal communication (email), 1 August 2003.

7. The eight countries were Australia, Brazil, Canada, France, Germany, Iran, Malaysia, and Nigeria.

8. Bill Mankin, Global Forest Policy Project, personal communication (email), 1 August 2003.

9. The IUCN is often seen as a NGO. In fact it is not a "pure" NGO, since membership of its decision-making bodies comprises both governments and NGOs.

10. For a thorough history and analysis of the ITTO, see Poore (2003).

11. The 140 NGOs originate from around the world as follows: 5 international NGOs; 39 Europe; 6 Africa; 7 Australia/New Zealand; 67 North America; 6 South America; 10 Asia.

8

Reflections on the Analytical Framework and NGO Diplomacy

Michele M. Betsill

In this chapter we return to the project's two main objectives: (1) to develop methodologies for strengthening claims about NGO influence in international environmental negotiations, and (2) through comparative analysis, to identify a set of factors that condition the ability of NGOs to influence these negotiations. We begin by reflecting on the first objective drawing on the empirical chapters presented in this volume. We argue that the framework succeeds in enabling scholars to make more robust claims about NGO influence, particularly by clarifying the links between NGO diplomacy and observed effects on negotiating processes and outcomes. The cases also demonstrate that it is possible to make qualitative judgments about levels of NGO influence and that such judgments provide a foundation for comparative analysis. Throughout the discussion we address some of the limitations of the framework in assessing the influence of NGO diplomats in any given set of negotiations. In the second section, we discuss the comparative analysis of the case studies. We address eight conditioning factors that emerged from the cases in this volume and consider whether these factors explain variation in NGO influence in international environmental negotiations. In the final section, we discuss the broader implications of this project. We examine how the present study relates to debates about NGOs and the effectiveness of environmental policy making, the democratization of global governance, and the changing nature of diplomacy in world politics.

Evaluating the Analytical Framework

In the introductory chapters it was argued that research on NGO influence in international environmental politics can be strengthened by recognizing more carefully the distinct political arenas in which NGOs operate, by defining what is meant by NGO "influence," and by elaborating the causal links between NGO participation and influence. In chapter 2 an analytical framework was presented for assessing NGO influence in one sphere of activity—international environmental negotiations. The framework, which begins with a definition of NGO influence as occurring *when one actor intentionally communicates to another so as to alter the latter's behavior from what would have occurred otherwise*, encourages scholars to draw on multiple types of data and analytical tools to evaluate the effects of NGO diplomats on both negotiation processes and outcomes. In the past, scholars often considered whether the effects of NGO activity can be seen in the agreement text. The cases analyzed in this volume clearly demonstrate that even where NGOs do not directly shape the final text, their influence is evident "behind the scenes" in terms of issue framing, agenda-setting, and/or shaping the positions of key states. In other words, when scholars focus only on the final outcome of negotiations in assessing NGO influence, they likely miss most of where NGOs "matter."

At the same time, although NGOs influence negotiations in multiple ways, that does not mean that their influence is everywhere. As we expand the types of claims we make about NGO influence, it becomes all the more important to support those claims through systematic analysis and to clearly identify instances of NGO successes *and failures*. The empirical chapters in this volume demonstrate the utility of our approach; assessments of NGO influence can be strengthened through systematic analysis involving triangulation of data sources and methods of analysis. Moreover the framework enables scholars to make qualitative assessments of NGO influence, which in turn provide a basis for conducting systematic comparative analyses of NGO influence in international environmental negotiations.

More Robust Claims

The empirical chapters in this volume demonstrate how our analytical framework can be used to make more robust claims about NGO influence. Our contributors drew on data obtained through participant observation, interviews, and/or negotiation documents to identify what NGO diplomats did at negotiating sessions (both formal and informal), what they hoped to achieve, and their effects on negotiation processes and outcomes. The case authors used these data to link NGO activities with observed effects through process tracing and/or counterfactual analysis, thereby strengthening claims about NGO influence. For example, Humphreys used process tracing to connect the dots from an NGO statement to the final text using negotiating documents and interviews in his analysis of ENGO influence on the negotiations related to the Intergovernmental Panel on Forests. In some instances our contributors used counterfactual thought experiments to identify NGO influence where it was not immediately obvious (e.g., the climate change case). Perhaps more important, counterfactuals also uncovered instances where NGOs were not influential despite a correlation between NGO activities and observed effects. In the whaling case, for example, Andresen and Skodvin questioned the influence of scientists in the International Whaling Commission's decision to adopt a new management procedure, noting that several other factors such as the rapid depletion of whale stocks and reduced profitability of the whaling industry made such a decision likely anyway. Similarly Humphreys challenged the claim that ENGOs convinced the World Trade Organization (WTO) to drop the Forests Products Agreement by identifying alternative explanations that would have led to the same result, even in the absence of NGOs.

Several project participants commented on the high data demands for using our approach. As discussed in chapter 2, we encourage analysts to gather data by attending negotiations sessions, reviewing all relevant documents, and conducting interviews with participants. Ultimately the framework requires that scholars have an intimate knowledge of the negotiating process they are studying, which has prompted numerous discussions about whether attendance at negotiating sessions is necessary. There are several benefits to attending negotiations. Scholars can obtain first-hand knowledge of the dynamics among participants and make important contacts. Several of our contributors attended most, if not all, of the negotiating sessions they reviewed. For instance, Corell and Burgiel attended virtually all of the Convention to Combat Desertification (CCD) and Cartagena Protocol sessions as reporters for the *Earth Negotiations Bulletins*, which also gave them behind the scenes access to the negotiations. Of course, it is not always possible for analysts to attend negotiations, often because of scarce resources but also because they are conducting an historical analysis where the sessions have already taken place. Betsill was only able to attend one session of the Kyoto Protocol negotiations, and found the experience enormously helpful in understanding the dynamics of the process. She augmented that knowledge by asking participants to describe what happened at other negotiating sessions in great detail during interviews. For those unable to attend the negotiations, the key is to ask, "What did I miss by not being there?" and to look for ways to fill the gaps using other data sources and/or analytical techniques.

The approach gives analysts some flexibility in assessing NGO influence, as illustrated by the fact that our contributors often used the framework in very different ways. Some followed the approach outlined in chapter 2 very closely, both in terms of data collection/methods of analysis and in the use of the indicators table to make qualitative judgments about levels of influence in a given case. Others applied the framework more loosely, selectively using particular aspects of the approach, such as process tracing, counterfactual analysis, and/or some, if not all, of the influence indicators in their final assessment. A few project participants felt the framework was too rigid, and they argued for adjustments in order to highlight what they viewed as the unique dimensions of their individual cases. The empirical chapters suggest that adjustments in individual cases can be made in some instances without compromising the framework's overall focus on systematic analysis. Moreover some of the adjustments were useful in highlighting broader issues related to NGO influence. For instance, Andresen and Skodvin's finding of ENGO influence through domestic channels in the whaling moratorium case prompted many useful discussions about the relationship between domestic and international politics in all cases, and ultimately some revision of the framework as noted in chapter 2.

Another difference in how the framework was applied in individual cases related to where scholars began their assessment of NGO influence. Betsill and Corell identified NGOs' self-defined objectives and considered whether they achieved those goals in the Kyoto Protocol and CCD negotiations. In contrast, Burgeil identified several of the central issues in the Cartagena Protocol negotiations and then evaluated NGO influence on each of the issues. Both approaches illuminated interesting aspects of the respective cases, and we do not wish to take a position on how things should be done in the future. However, we do want to highlight the implications of making this choice. In the first approach, where the focus is on whether NGOs achieve their own goals, scholars may be inclined to identify instances in which NGOs failed to exert influence (e.g., where NGOs did not achieve a particular goal). At the same time scholars run the risk of overestimating the overall influence of NGO diplomats in a particular set of negotiations. We can imagine a case where NGOs achieved all their goals (suggesting a high level of influence) but on relatively marginal issues. Adding to these difficulties, Giugni (1999) notes that assessments of goal attainment are often subjective, with participants disagreeing whether a goal has been achieved. In addition he argues that focusing on goal attainment overemphasizes intention and ignores the possibility that observed effects may be unintended consequences of NGO action. Focusing on the major issues under negotiation may better tell us the larger impact of NGOs in a set of negotiations (e.g., they were/were not influential on the major issues) but give us an incomplete picture of NGO influence. Such an approach assumes NGOs were focusing their efforts on these issues, so scholars should be careful to distinguish between findings of low influence (where NGOs were trying to influence an issue but failed) and no influence (where NGOs were inactive on an issue). Also some instances of NGO success may be overlooked in this approach.

There were limits, however, to how far the framework can stretch. The framework is easiest to apply to cases focused on the negotiation of specific agreements (the Kyoto Protocol, CCD, and Cartagena Protocol cases), and we are relatively confident in the assessments of NGO influence in these cases. Applying the framework becomes more difficult when analyzing multiple negotiations in an issue area over time (e.g., the whaling and forestry cases). As the data demands become more cumbersome, we suspect that assessments of NGO influence are overdetermined as data are aggregated over longer periods of time. For example, Humphreys found a high level of NGO influence in his analysis of forestry negotiations between 1983 and 2000 despite the many instances where ENGOs failed to achieve their objectives.

A way to avoid this may be to assess the influence of NGO diplomats in discrete sets of international environmental negotiations. As discussed in chapter 1, we chose to focus on this relatively narrow area of NGO activity in global environmental politics so that we could generate some theoretically useful findings, but we recognize that our decision comes with trade-offs. Our framework was developed with the political arena of multilateral negotiations in mind (e.g., we chose a definition of influence specific to this context). We therefore urge scholars to exercise caution and make necessary adjustments when applying the framework to other contexts (e.g., domestic politics and/or intra-NGO relations in the realm of global civil society). We expect our approach to be of greatest use to scholars coming from a liberal institutionalist or pluralist perspective who wish to assess the influence of NGOs on the development of specific policies. Presumably many NGOs participate in international environmental negotiations in order to shape their outcomes, so it is useful to explore the question of when they are more or less likely to do so.[1] However, our approach may be less useful for constructivists interested in identifying the ways that NGOs contribute to broader changes in ideas and issue framing, since this process takes place over a longer period of time and often in multiple institutional settings. As Humphreys argues (see also Humphreys 2004), ENGOs' most significant impact on international forestry politics has been in altering how the issue has been framed over time rather than having specific text adopted in any given agreement. We will let others debate which of these political arenas and processes are most important in the grand scheme of things. In the meantime, we urge scholars who use this framework to be cognizant of the fact that international negotiations are just one of many political arenas in which NGOs engage in global environmental politics and that any given set of negotiations takes place in a larger historical and ideational context.

While we are pleased to see the multiple ways in which our approach can be used to analyze individual cases of NGO diplomacy, we have argued for going beyond identifying the unique aspects of these cases if we are to advance our understanding of NGO influence in international environmental negotiations. As discussed in chapter 2, our framework is designed to facilitate structured, focused comparison across cases by identifying a set of general questions that can be asked in each case about the link between NGO activities and observed effects, particularly related to the negotiation process (issue framing, agenda-setting, and the positions of key states) and outcome (on both procedural and substantive issues) (George and Bennett 2005). When some of our contributors argued that the framework was too rigid, we suspect they were reacting to our demands for structure and focus rather than the nature of the framework itself.[2] Those who use the approach in the future will probably need to establish their priority—explanation of an individual case or comparison across cases—before deciding exactly how to apply the framework.

Qualitative Assessments of Influence

The cases demonstrate that it is possible to make qualitative judgments about levels of NGO influence in international environmental negotiations. The framework allows us to differentiate among at least three levels of NGO influence. Where NGOs actively participate in the negotiations but have virtually no observable effect on the negotiation process or outcome, we can say that they have exerted a low level of influence. In the five case chapters presented here we did not have any instances of low-level influence. However, as discussed below, we disaggregated some of these contributions for the comparative analysis and found two cases of low influence: ENGOs in the United Nations Forum on Forests negotiations and scientists on the moratorium decision in the International Whaling Commission (see table 8.1). Moderate influence occurs when NGOs actively engage in the negotiations and have observable effects on the negotiating process but not on the final outcome. The Kyoto Protocol was the clearest example of this case. The ENGOs shaped the negotiating agenda on emissions trading and sinks, and without them the United States may not have given in on reduction targets. However, the final treaty does not reflect any of the ENGO positions. Finally, NGOs have achieved high levels of influence when their activities are linked to observable effects on both negotiation process and outcome, as occurred in the CCD.

While the framework does allow for a differentiation in the levels of influence, it was not always straightforward which category was most appropriate in any given case. For example, in the biosafety case both ENGOs and industry groups influenced the negotiation process and achieved their goals on some, if not all, issues in the final treaty. Humphreys found several instances in which versions of proposals that had been introduced by NGOs were included in treaty text, but only after being diluted by state delegates. In all these instances the final assessment of NGO influence could not be easily categorized as moderate or high. In the comparative analysis below we added a fourth category: moderate/high indicating that NGO diplomats had achieved some influence on negotiating processes and outcomes but had not achieved all of their goals (again, distinguishing these cases from the CCD). Scholars interested in using this approach may wish to refine these categories even further.

It may be easier, and perhaps more appropriate, to assess the influence of NGO diplomats at the level of individual issues under negotiation rather than on the overall negotiations. Negotiation processes related to the environment are highly complex and can cover numerous technical issues simultaneously. Comparative analysis could examine the relationship between levels of NGO influence and the nature of the different types of issues under negotiation. For example, it may make a difference whether an issue involves positional or distributive bargaining, where the key question is who should reap the benefits/incur the costs of a particular policy.[3] Such analyses would likely result in a larger number of cases (since multiple issues may be addressed within a single set of negotiations) and would allow scholars to control some of the conditioning factors (e.g., presence of other NGOs). In practice, linkages between issues make it difficult to treat issue negotiations separately, as became evident in the Kyoto Protocol negotiations where debates over targets and timetables were connected to discussions over flexible mechanisms. But this is just one example of how scholars using the framework in the future might consider alternative ways of defining and determining levels of influence that arise during multilateral negotiations.

Ultimately we were interested in generating qualitative assessments of NGO influence to facilitate comparison across cases. While the existing literature clearly shows that NGOs influence international environmental negotiations in various ways, further theoretical development requires moving beyond to the question of *the conditions* under which NGOs' influence matters. The systematic approach achieved through the use of the framework has illuminated *variation* in NGO influence across cases. Such variation allowed us to assess of a set of conditioning factors that shape the ability of NGO diplomats to influence multilateral negotiations on the environment and sustainable development. The next section reports the findings that emerged across the analyses in this volume. To some extent, each set of negotiations was found to be unique in that NGO diplomats seeking to influence a particular set of negotiations must be able to recognize and adapt to its distinctive features (i.e., no "magic bullet" could be found for NGO influence). At the same time it does appear that some factors make NGO influence more or less likely across negotiating situations.

Comparing across Cases: Conditioning Factors

In order to complete this comparative analysis, it was necessary to make some adjustments to the cases analyzed in this volume. First, some of our case studies had to be disaggregated once we realized the framework was more useful in analyzing discrete sets of negotiations (table 8.1). As a result Burgiel's case study had to be treated as two separate subcases: one on the influence of ENGOs and one on the influence of industry on the Cartagena Protocol negotiations. Andresen and Skodvin's whaling chapter was separated into three subcases,[4] and Humphreys' forest chapter was divided into nine subcases. In some instances rough assessments were made of levels of NGO influence that drew on material presented in the respective chapters and used the analytical framework introduced in chapter 2.[5] Because the authors themselves had not conducted the analyses in this manner, it was difficult to collect all of the data we needed to have fully comparable cases. As noted above, we added a category of "moderate/high" influence where NGO diplomats shaped the process and were successful in influencing some but not all aspects of the outcome (our assessment of NGO influence in the biosafety case thus differs from Burgiel's). Disaggregating in this way provided additional

Table 8.1
Cases for the comparative analysis

Negotiation	Agreement year	Type of NGOs	Level of influence
Kyoto Protocol	1997	ENGOs	Moderate
Cartagena Protocol on Biosafety 1	2000	ENGOs	Moderate/high
Cartagena Protocol on Biosafety 2	2000	Industry	Moderate/high
UN Convention to Combat Desertification	1994	ENGOs/social NGOs	High
International Whaling Commission New Management Procedure	1974	Scientists	Moderate
International Whaling Commission Moratorium 1	1982	ENGOs	High
International Whaling Commission Moratorium 2	1982	Scientists	Low
UNCED Forest Principles	1992	ENGOs	High
Intergovernmental Panel on Forests	1997	ENGOs	Moderate/high
Intergovernmental Forum on Forests	2000	ENGOs	Moderate/high
United Nations Forum on Forests	2000	ENGOs	Low
International Tropical Timber Agreement 1	1983	ENGOs	High
International Tropical Timber Organization—labeling	1989	ENGOs	Moderate
International Tropical Timber Organization—sustainable forest management	1990	ENGOs	Moderate
International Tropical Timber Agreement 2	1994	ENGOs	Moderate
World Trade Organization Forest Products Agreement	1999[a]	ENGOs	Moderate

a. This case is a "nonagreement."

leverage in the comparative analysis by expanding the number of observations and, to a limited degree, variation on the dependent variable, NGO influence (King, Keohane, and Verba 1994).

To identify the set of conditioning factors for further analysis, an inductive process was used for each of the case studies. We asked the authors to identify what they viewed as the key factors that enhanced or constrained the ability of NGO diplomats to influence international environmental negotiations. Eight factors came up most frequently across all of the cases: (1) NGO coordination, (2) rules of access, (3) stage of the negotiations, (4) political stakes, (5) institutional overlap, (6) competition from other NGOs, (7) alliances with key states, and (8) level of contention. This should not been seen as an exhaustive list of the factors that might shape the ability of NGOs to influence international environmental negotiations. The general literature on NGOs (see chapter 2) as well as several of our case studies suggest many others that need to be analyzed more systematically.

As discussed in chapter 2, discussions of conditioning factors often distinguish between agency and structure in explaining variation in NGO influence across negotiations. Agent-based factors emphasize the behavior and/or characteristics of NGOs and imply that NGO diplomats can make choices to enhance their influence. Only one of the eight factors identified in our cases (NGO coordination) relates to agency. The remaining factors are structural in that they point to the importance of context and suggest that NGO diplomats are enabled or constrained by elements of the setting in which negotiations take place. Structural factors help explain why NGO diplomats have different levels of influence across negotiations despite employing similar strategies. Six of the seven structural factors relate to institutional elements of the structure, comprising what social movement scholars refer to as the political opportunity structure, which is characterized by the formal organizational/legal structure and power relations between actors participating in the negotiations (McAdam 1996). Rather than construct a single measure of political opportunity structure, we find it more useful to think of political opportunity structures as clusters of variables and to analyze whether and how specific aspects of the institutional context shape NGO opportunities for influence (see Gamson and Meyer 1996). The final conditioning factor (level of contention) highlights a cultural element of the negotiating structure related to issue framing.

In table 8.2 we summarize these factors. In the discussion below we consider their explanatory value and discuss their connection to the broader literature on NGOs in international environmental negotiations and global environmental politics. Readers should use caution in generalizing these findings beyond our cases for a number of reasons. This exercise is best viewed as a "plausibility probe," suggesting potential avenues for future research, rather than a formal "test" of the factors given the limitations of our approach to case selection. We selected cases based on the availability of scholars with prior knowledge of NGO diplomacy and made no determination on the appropriateness of the cases at the outset for analyzing specific propositions related to conditioning factors. In addition the majority of our cases examine ENGOs, so we were limited in what we could say about differences in the conditions under which different types of NGOs would influence international environmental negotiations. Nevertheless, to make our observations about possible differences, we clearly required a more systematic approach to reach any strong conclusions. Finally, more than half of our cases focus on forestry issues, creating both challenges and opportunities in conducting our comparative analysis. On the one hand, we had to be careful not to generalize too heavily from our cases to other areas of environmental politics. On the other hand, we could hold some things constant in order to evaluate the relative importance of other conditioning factors. We strongly encourage scholars to subject the issues raised here as well as propositions from the broader literature to rigorous analysis based on a more careful selection of cases. However, despite these reservations, we suspect that many of our findings will hold up in other multilateral negotiation settings, including those outside the environmental issue area.

NGO Coordination

Several contributors highlighted coordination between like-minded NGOs as a strategy that enhanced the ability of NGOs to influence international environmental negotiations. In the Kyoto Protocol case, Betsill argued that members of the Climate Action Network were able to influence the negotiation process by speaking with one voice. Corell made a

Table 8.2
Summary of findings

NGO coordination
• Neutral effect

Rules of access
• NGO influence does not decline when rules become more restrictive
• NGO influence is enhanced when active steps are taken to facilitate NGO participation

Stage of negotiations
• NGO influence is more difficult during the detail phase of negotiations
• For ENGOs, influence during the formula phase is necessary, but not sufficient, to achieve influence during the detail phase

Political stakes
• High levels of NGO influence are most likely when political stakes are low
• As political stakes increase, NGO influence is enhanced when delegates see NGOs as trusted partners in achieving objectives

Institutional overlap
• NGOs can influence negotiations indirectly by influencing related institutions
• Overlap with the WTO and international trade regime constrains ENGOs and enables NGOs representing business/industry interests

Competition from other NGOs
• NGO influence is not necessarily constrained when there is competition from other NGOs (NGO influence is not a zero-sum game)

Alliances with key states
• NGO influence is enhanced when they form alliances with key states
• Alliances with key states have limited utility when there is a high degree of polarization between states

Level of contention
• ENGO influence is constrained where entrenched economic interests are at stake
• Influence of NGOs representing business/industry interests is enhanced when entrenched economic interests are at stake

---

similar argument in the desertification case. In both instances, environmental (and social) NGOs adopted an explicit strategy of coordinating their messages and lobbying activities. In the International Whaling Commission revised management procedure case, Andresen and Skodvin also identified coordination as an important conditioning factor but in a slightly different way, focusing on the coordination of scientists' beliefs (scientific consensus) rather than their messages and strategies (see also Haas 1992). Many scholars have argued that coordination between NGOs in negotiation situations results in greater efficiency, which in turn should enhance the ability of NGO diplomats to exert influence (e.g., Betsill 2002; Corell and Betsill 2001; Dodds 2001; Duwe 2001; Biliouri 1999; Keck and Sikkink 1998; Chatterjee and Finger 1994). However, among our cases, coordination had a neutral effect on NGO influence (Arts 2001 arrived at a similar conclusion). NGOs attained all levels of influence under conditions of coordination. Of particular note, ENGOs achieved a high level of influence in the UNCED Forest Principles negotiations, even though their coordination was loose and they had no unified position on the need for a treaty.

While coordination might not have been an important factor related to NGO influence, this discussion and our cases call attention to the highly political nature of inter- and intra-NGO relations. Like states, NGOs are political actors, with their own power relations and contentious internal debates (see also Friedman, Hochstetler, and Clark 2005; Hochstetler 2002; Carpenter 2001; Duwe 2001; Jordan and Van Tuijl 2000; Chatterjee and Finger 1994). For example, Andresen and Skodvin discussed polarization within the scientific community during the whaling moratorium debates, and Betsill noted deep cleavages along North–South lines among members of the Climate Action Network during the Kyoto Protocol negotiations. Even among NGOs with seemingly common interests, arriving at a consensus position is frequently mired in controversy linked to inequalities between large, well-funded international NGOs and smaller grassroots organizations or different ideas about how global environmental problems should be addressed (Friedman, Hochstetler, and Clark 2005; Duwe 2001). We return to this point below in our discussion of NGOs and democratizing global governance.

Rules of Access

There are no set rules governing NGO participation in international environmental negotiations. The tendency has been for the international organizations responsible for a particular negotiation to establish rules for NGO access, often on an ad hoc basis, and there is a great deal of variation among international bodies (International Centre for Trade and Sustainable Development 1999). Raustiala (1997, 2001) notes that states usually determine these rules, implying that the power to do so is a source of leverage for states over NGOs. Indeed several of the individual case authors argued that the access rules applicable in a particular negotiating context shaped the ability of NGO diplomats to exert influence. Specifically, the more restrictive the access, the less influence NGOs were assumed to have.

The relationship between access and influence became more complex when we examined the issue on a comparative basis. On the one hand, we found two instances in which steps were taken to reach out to NGOs and actively facilitate their involvement in the negotiating process. In the CCD negotiations, the Secretariat and chair of the negotiating body were committed to ensuring that NGO diplomats had full and open access to documents and were permitted to make oral interventions and distribute statements on more or less the same basis as states. In the case of the International Whaling Commission's new management procedure, there was a formal institutional space—the Scientific Committee—through which scientists could promote their agenda. In both cases, states were particularly dependent on non-state actors to provide specific expertise. In other words, NGOs were seen as important partners in helping states achieve their interests (Arts 2001; Dodds 2001; Kellow 2000; Raustiala 1997). States then were willing to adopt positive access measures, which in turn likely enhanced the ability of NGO diplomats to influence the negotiations.

In many other cases, states tried to restrict NGO access to the negotiations. However, we did not find that a more restrictive environment for NGO access necessarily constrained the ability of NGO diplomats to influence international environmental negotiations. Time and again NGO diplomats overcame restrictions on their access by using alternative strategies. For example, when the rules for NGO access became more restrictive in the Kyoto Protocol and Cartagena Protocol negotiations, NGO diplomats used cell phones, relied heavily on personal contacts with delegates, and/or managed to secure positions on state delegations and were subsequently able to exert some degree of influence. Negative access measures clearly did not constrain the ability of NGOs to influence the negotiations.

These findings are interesting in light of the fact that NGOs expend considerable time, energy and resources trying to secure open access to international decision-making processes. NGO coalitions, such as the Access Initiative and the Public Participation Campaign of the European Eco Forum, have supported the negotiation of the Aarhus Convention on Access to Information, Public Participation in Decision-making and Access to Justice in Environmental Matters and have engaged in discussions to enhance NGO access in institutions such as the United Nations Environment Programme and the Commission on Sustainable Development (European Eco Forum 2005; The Access Initiative 2005; Dodds 2001; UNEP 2001). These activities are based on an assumption that by securing official opportunities for access to decision-making processes, NGOs can enhance their ability to influence such processes (Corell and Betsill 2003; Dodds 2001). Similarly states seem to equate NGO access with influence, and thus they routinely invoke their sovereign privilege to restrict official NGO participation in negotiations (Oberthür et al. 2002; UNEP 2001; Clark, Friedman, and Hochstetler 1998). Our findings suggest that NGOs must do more than simply force states to open up official avenues for participation if they hope to enhance their influence. NGO diplomats need to convince state decision makers and international organization officials that they can be effective partners in helping negotiators make better decisions and/or implement those decisions. Then state delegates may be willing to take active steps to facilitate (rather than merely allow) the participation of non-state actors.

Stage of Negotiations

Our cases highlight a temporal dimension of the institutional structure and suggest a potential link between the negotiation stage and NGO influence. Specifically, it may be analytically useful to differentiate between distinct, but overlapping, phases of negotiation processes. For example, Zartman and Berman (1982) distinguish between a formula phase where participants agree upon a framework for the negotiations and a detail phase where they bargain over the specifics of the final text (see also Chasek 2001). In all but two cases we found that NGOs exerted influence in the formula phase of negotiations, especially through shaping the negotiating agenda. For instance, in the Kyoto Protocol negotiations, ENGOs were credited with shaping debate on emissions trading and the inclusion of sinks. The exceptions were industry in the biosafety negotiations and ENGOs in the United Nations Forum on Forests case. In our cases, NGOs had less success in influencing negotiating processes and outcomes as the talks moved into later stages where positions had hardened and states had to resolve core issues. As Humphreys notes in the case of forests negotiations, there may be less "political space" available to NGO diplomats at the detail phase of negotiations. Similarly Burgiel argues that debates during the later stages are more heavily politicized. More pragmatically, this may reflect a tendency to simplify negotiations at the final stages by reducing the number of people in the room (Corell and Betsill 2003).

Many scholars have argued that NGOs are most influential during the agenda-setting phase of international policy making (prior to the negotiation phase) where they catalyze action by identifying problems and calling upon states to act (Gemmill and Bamidele-Izu 2002; Raustiala 2001; Yamin 2001; Newell 2000; Biliouri 1999; Charnovitz 1997).[6] Our findings disagree in one sense. We clearly find that NGO diplomats can be influential during the policy formulation/negotiation phase of international policy making as well. However, our findings do reinforce the idea that NGOs have the greatest effect on agenda-setting, particularly if we think of agenda-setting as an ongoing process rather than a distinct stage of policy making that ends once negotiations begin.

Here we found an interesting difference between ENGOs and industry groups. In our cases, instances of ENGO influence over the final text were *always* preceded by influence on the negotiating agenda. Where ENGOs failed to shape the agenda (e.g., in the United Nations Forum on Forests negotiations), they also failed to influence the negotiation outcome. Our cases suggest that for environmentalists, influence in the early stages of negotiations may be necessary (though by no means sufficient) for achieving influence in later stages. This pattern did not hold, however, in our one case involving industry. In the biosafety negotiations, industry groups had little influence on setting the negotiating agenda, and yet they were able to influence the outcome of the negotiations on several issues. Burgiel explains this by highlighting the fact that ENGOs and industry groups had different objectives. Industry groups were primarily concerned with limiting the protocol's scope and keeping some issue off the table. In other words, they were less concerned with getting things on the agenda than with taking them off, which can be accomplished at all stages of the negotiations. In contrast, ENGOs are often most interested in getting (and keeping) issues on the negotiating agenda. As suggested above, success may require achieving influence early on, since it becomes much harder to get an issue on the agenda once the negotiations move into their detail phase.

Political Stakes

Another element of the political opportunity structure concerns the legal nature of the negotiations. Several of our contributors highlighted the importance of the political stakes in shaping NGO influence. This points to a slightly different temporal dimension of international environmental negotiations, reminding us that treaty regimes evolve over time (see Spector and Zartman 2003; Tolba 1998; Széll 1993; Zartman 1993). Initial agreements (sometimes referred to as "framework" agreements or declarations) often articulate general principles, establish new organizations and/or decision-making procedures, but may not require significant behavioral change from member states. Post-agreement negotiations focus on how to achieve treaty goals and address ongoing or new conflicts that arise (Spector and Zartman 2003). In some cases this may involve negotiating a new instrument (usually referred to as a "protocol") or creating new institutions and decision-making procedures to enhance the implementation of the initial agreement. Post-agreement negotiations typically address more substantive issues and are more likely to seek specific behavioral changes from member states.

One could argue that for states, the political stakes are lower in the negotiation of an initial agreement on a given issue, which typically involves debates on general principles and institutional procedures rather than specific behavioral commitments. As a result states may be more willing to give NGO proposals serious consideration and perhaps to make concessions. Two of our three instances of high NGO influence fall into this category. The CCD was a framework agreement, and the 1983 International Tropical Timber Agreement set up a new organization. By this same logic, we might expect NGO influence to decrease during post-agreement negotiations as the political stakes increase. When states are negotiating more specific behavioral commitments, one might expect them to protect their interests more strongly and be less willing to consider and/or adopt NGO proposals. In our cases, NGO diplomats seemed to have a harder time exerting influence when negotiations moved beyond the initial agreement stage. We clearly saw this trend in our forestry cases, which covered a period from 1983 to 2000 and where we observed the full range of NGO influence. The case of high influence (1983 ITTA) occurred in the earliest set of negotiations, again, where states were debating general principles to guide behavior on forestry issues, and the low influence occurred in the most recent round of negotiations (United Nations Forum on Forests) where states were being asked to implement specific policy changes. In the forestry case this trend held across institutional contexts, suggesting this may be an especially strong conditioning factor.

At the same time it is notable that NGO influence did not disappear altogether in post-agreement negotiations. In fact we had several instances of moderate/high levels of influence in post-agreement negotiations, and one of our high levels of influence occurred in the whaling moratorium case. The classic explanation is that prospects for cooperation increase when individuals interact repeatedly (Axelrod 1984). Cooperation between states is thought to become possible when they engage in multiple rounds of negotiations on a particular issue over time. The same may be said for state–NGO interactions. When the political stakes increase for states, they may be more willing to work with NGO diplomats if they already have had experience doing so. Several contributors noted that NGO representatives were well respected by state delegates and that the close personal relationships that develop over time can open up opportunities for NGO influence.

States have been known to rely more heavily on some NGOs when negotiations move to the post-agreement stage, thereby presenting further opportunities for influence. Negotiations over particular regulations or commitments can be extremely technical, requiring expertise that state delegates may not possess themselves. Many of our contributors spoke of NGOs' specialized knowledge as an important source of leverage in international negotiations. States looking for viable policy options may turn to NGO diplomats for information about the potential costs and benefits of particular policies or for technical information about how a policy might be implemented. In other words, NGO influence in post-agreement negotiations, where political stakes are higher, may be attributable to the fact that states need NGOs to achieve their objectives (Raustiala 1997). Of course, the advantage only goes to those NGOs with the required expertise.

Institutional Overlap

Our cases confirm the fact that international environmental negotiations take place in an increasingly dense network of overlapping regimes and institutions. Such overlap can both enable and constrain the ability of NGOs to exert influence (see also Selin and VanDeveer 2003; Rosendal 2001a). Decisions made in one negotiating context can directly affect the political opportunities available to NGOs in other contexts. For example, the desertification negotiations were much more open to NGO participation than many of the other cases examined, in large part because of the precedent set at the 1992 United Nations Conference on Environment and Development in Rio de Janeiro. Many negotiating processes that began prior to the Rio Summit subsequently revised their rules to open up opportunities for NGO participation. Similarly ENGOs successfully campaigned to ensure the rights of local communities, indigenous peoples, and women to participate in forest policy in the UNCED forest principles negotiations, so they were able to capitalize on this success when negotiations moved to a new institutional setting, the Commission on Sustainable Development.

This suggests that NGOs can influence a given negotiation process not only by participating in those negotiations but also by exerting influence in a related institutional setting. NGO diplomats may wish to engage in "venue shopping" and search for the institutional context where they are most likely to have influence (Alter and Meunier 2006). Interestingly related negotiations in different institutional settings are often unconnected in terms of the NGO communities that participate. For example, forest policy is debated in some settings specifically dedicated to forests (e.g., the United Nations Forum on Forests) as well as in broader contexts related to climate change, biodiversity, and desertification. Despite these linkages negotiations tend to involve distinct NGO communities that often fail to effectively communicate with one another (Corell and Betsill 2003).

Institutional overlap may limit NGO influence in some instances. In our cases this was true for ENGOs in the biosafety negotiations and the International Tropical Timber Organization talks on labeling where the issues under negotiation were closely linked to the WTO and the international trade regime. In each case ENGOs sought to limit states' economic activities, contrary to neoliberal economic norms, and opponents threatened to use the WTO as an alternative venue for promoting their interests.[7] ENGOs were thus limited in what they could demand and achieve in these cases. They recognized that their chances of influence would be much lower in the WTO, which environmentalists tend to view as hostile to their interests (see Williams and Ford 1999).

This discussion is interesting in light of current debates about creating a world environment organization and multilevel governance. Biermann (2000) argues that a world environment organization would make environmental policy making more efficient by coordinating international treaty regimes on disparate issues under one umbrella. This could lead to more standardized norms on decision-making procedures, a more streamlined set of negotiating bodies, and more explicit connections between issue areas. All NGOs might then acquire greater clarity in terms of rules of access, have fewer meetings to attend, and build up expertise at linking issues. Nevertheless, consolidating global environmental governance can limit the range of institutional options available to NGO diplomats to influence negotiations in a particular issue area.

In a parallel discussion, scholars increasingly acknowledge that environmental issues are governed at a variety of tiers and spheres of governance where authority is shared between state and non-state actors operating on the local, regional, national, transnational, and global scales (Betsill and Bulkeley 2006; Hooghe and Marks 2003; Vogler 2003; Young 2002). This multilevel governance perspective suggests that opportunities for NGOs to influence the governance of these issues extend beyond other multilateral negotiation processes. For example, when the World Wide Fund for Nature was unsuccessful in promoting timber labeling within the realm of formal intergovernmental politics, it took matters into its own hands and established the Forest Stewardship Council to work directly with timber companies.

Competition from Other NGOs

Power relations are another aspect of the political opportunity structure. The presence (or absence) of other NGOs is frequently noted as a factor shaping NGO influence in international environmental negotiations. Betsill reported that ENGOs involved in the Kyoto Protocol negotiations spent a great deal of time countering arguments made by the fossil-fuel industry, which strongly opposed international regulations on GHG emissions, taking time away from more direct efforts to promote their own agenda. Where NGOs promote competing interests (e.g., environmentalist against industrial interests), the assumption is that the groups will offset one another, making it difficult for either group to exert influence on the negotiations. It is striking that there were no other NGOs present in the CCD negotiations, and further only a weak presence of other NGOs in the whaling moratorium, both cases with high levels of NGO influence. However, other factors appear to have been more important in explaining the high levels of NGO influence. In the desertification case, NGO diplomats were given a privileged position in the negotiations (see access discussion above), and in the moratorium case, the framing of the whaling issue as a moral concern clearly privileged the ENGO position over the scientific position. In addition scientists were divided on the desirability of a moratorium.

In several cases where other NGOs were present, NGO diplomats were still able to achieve moderate and moderate/high levels of influence, suggesting that the presence of competition is not necessarily as constraining as one might expect. The biosafety case is particularly illuminating as it shows that NGO influence in international negotiations is not a zero-sum game. In this case environmentalists and industry directly opposed one another on some issues but otherwise pursued separate agendas. Where they pursued separate issues, each was able to achieve some success without necessarily taking anything away from the other. In comparing two types of NGOs in a single case, Burgiel shows us not only how different groups use different strategies but also how the institutional context creates different types of political opportunities for these groups (we also see this in Andresen and Skodvin's whaling case). Similarly, in an analysis of the climate change negotiations, Newell (2000) found that different types of NGOs (media, scientific, industry, and environmental) exerted different types of influence. Future research comparing the influence of different NGOs in a single set of negotiations would advance our ability to draw generalizable conclusions about the significance of these differences. Moreover it would be useful for analysts to take a much more sophisticated approach to the study of multilateral negotiations by considering how NGO diplomats interact with one another (as well as states) and with what effect (see Rowlands 2001).

Alliances with Key States

In all our cases the abilities of NGOs to influence international environmental negotiations improved through alliances made with key states. NGOs often shaped the position of a key state or group of states by using domestic and international channels. In the whaling moratorium case, ENGOs mobilized public opinion in the United States and subsequently shaped the American government's position. At the international level, ENGOs used their powerful US ally to generate pressure on whaling states to change their behavior. Similarly ENGOs in the Kyoto Protocol negotiations employed domestic channels to develop an alliance with the European Union, which in turn helped them pressure the United States to change its position on emissions targets (see also DeSombre 2000). Even where ENGOs did not shape state positions per se, their ability to influence negotiations was enhanced when they put forward proposals that resonated with the interests of key states, as in the Intergovernmental Panel on Forests case and the protection of traditional forest related knowledge (see also Hochstetler 2002; Arts 2001). In the CCD negotiations, Corell argued that NGOs benefited from an alliance with donor countries, who saw NGOs as agents of economic and political change in developing countries.

Our cases also suggest that the utility of NGO–state alliances depend on the general relationship between states in the negotiations. Many of the forestry negotiations were highly polarized along North–South and producer–consumer lines. Neither side was sufficiently strong to dominate, and this limited the significance of alliances for ENGOs. In the International Tropical Timber Organization negotiations on trade in forest products, ENGOs successfully convinced the British to table their proposal on labeling, but because support was lacking from producer states, it was not possible for the proposal to be adopted. Arts (2001: 208) argues that "whenever North and South really clash on environmental matters, the situation becomes so heavily politicised that intervention by any third party is doomed to fail." While we did find polarization to be a constraining factor, we would not go so far. In debates over the whaling moratorium, states were highly polarized, but the membership of the International Whaling Commission was heavily weighted toward non-whaling states, making ENGOs' alliance with the United States particularly effective.

This discussion highlights the fact that NGOs and states interact in complex ways in multilateral negotiations. Scholars and practitioners typically assume that NGOs and states compete against one another in international forums (e.g., Newell 2000; Close 1998; Willetts 1996b). While such competition is often the case, our case studies suggest that NGOs and states also frequently work in alliance with one another (see also Falkner 2003; Gulbrandsen and Andresen 2004). For example, business NGOs capitalized on their relationships with members of the Miami Group during the Cartagena Protocol negotiations to promote their preferred positions. In negotiating the CCD, southern NGOs worked closely with northern governments, who wanted to bypass corrupt regimes and work directly with NGOs in developing countries. Such relationships may be developed at the international level during the negotiations or, as in the case of the negotiations on whaling, when delegates return home in between formal negotiating sessions. Friedman, Hochstetler, and Clark (2005) elaborate on this complexity in their analysis of state–

NGO interactions in UN-sponsored world conferences, particularly over issues of sovereignty. Over the years, states have opened up new avenues for NGO diplomacy in negotiating processes, leading some to presume that NGOs are infringing on state sovereignty (Mathews 1997). At the same time states have often pushed back and placed limits on NGO participation at crucial moments in the negotiation process. Friedman and colleagues highlight the numerous "sovereignty bargains" that characterize state–society relations in multilateral negotiations.

Level of Contention

A number of case authors identified the framing of the issue under consideration as a factor that shapes the ability of NGO diplomats to influence international environmental negotiations. Of particular import was the perceived level of contention, especially the extent to which negotiators understood there to be entrenched economic interests at stake. The CCD negotiations, where environmental and social NGOs had a high level of influence, showed the lowest level of contention over economic interests of all our cases. In contrast, ENGOs were clearly constrained in the Kyoto Protocol and the Cartagena Protocol negotiations because of the perceived links between these issues and the core economic activities of states. On climate change, decisions about limiting greenhouse gas emissions were understood to have implications for energy prices and industrial production, issues as the heart of industrialized countries' economies. As a result negotiators often ignored ENGO arguments about the long-term costs of inaction, focusing instead on the short-term costs of controlling emissions. Similarly ENGOs had difficulties promoting their agenda in the Cartagena Protocol negotiations, since it was largely framed as a trade issue (rather than a broader health issue) from the outset. Proposals to limit economic activity by regulating trade in genetically modified organisms went against the neoliberal economic norms governing international trade. Conversely, Burgiel argues that this framing may have enhanced the ability of industry groups to resist such regulations in the biosafety case.

Our findings are consistent with Bernstein's (2001) expectations about the political implications of the liberal environmental norms that dominate the current international system, which assume the compatibility of economic growth and environmental protection and accept the basic tenets of the market economy. He predicts, "policies that contradict key norms of liberal environmentalism are more likely to face strong contestation or not even be considered owing to the prevailing norm complex" (Bernstein 2001: 235). Indeed Humphreys noted in the forestry case that ENGOs had greater influence when their arguments were framed in terms consistent with a neoliberal economic discourse (see also Williams and Ford 1999).

Humphreys identified sovereignty as an area of contention in the forestry negotiations. Many states raised strong objections to clauses on the rights of indigenous peoples on that basis that this could constitute an erosion of sovereignty over natural resources within their territories. However, our analysis suggests that contention over sovereignty had a neutral overall effect on the influence of NGO diplomats in forestry negotiations. While it is doubtful that sovereignty concerns enhanced opportunities for NGO influence, NGO diplomats did succeed in having indigenous peoples' rights recognized in several forest-related treaties (the Intergovernmental Forum on Forests, the Intergovernmental Panel on Forests, and the UNCED Forest Principles), so sovereignty concerns were not necessarily constraining in this case.

Friedman, Hochstetler, and Clark's (2005) recent study on NGOs and world conferences found that states are particularly resistant to NGO proposals that are understood to challenge aspects of sovereignty, especially control over internal affairs and autonomy over national economic decisions. Proposals to protect the rights of indigenous peoples in forestry policy can be threatening to states on both counts, and yet NGO diplomats were able to get this issue included in the text of several agreements. Friedman and colleagues suggest one possible explanation: states, especially Southern states, were willing to make such "sovereignty bargains" in the mid-1990s with the expectation that doing so would generate new financial resources for development.[8] They contend that this changed following the Rio+5 meeting in 1997 when it became clear that new resources were not forthcoming. Many of Humphreys' cases in which NGOs were successful in promoting indigenous peoples rights occurred prior to this change.

Broader Implications

In addition to the methodological and theoretical contributions noted above, our case studies speak to broader debates in the fields of global environmental politics and international relations. They raise questions about the links between NGO influence and environmental outcomes, the democratization of global governance, and the changing nature of diplomacy in world politics. While our framework does not address these questions directly, the case studies contribute to these debates by illuminating the nature of NGO participation and influence in international environmental negotiations.

Improved Environmental Outcomes

It is often argued that increasing NGO participation in and influence on multilateral negotiations on the environment and sustainable development leads to "better" outcomes. One line of reasoning contends that NGOs improve decision making by providing valuable expertise (see Susskind et al. 2003; Dodds 2002; Corell 1999b). Our cases confirm that NGOs often help decision-makers navigate the highly complex and technical nature of many environmental issues. The logic here assumes that better information leads to better outcomes. Another argument is that NGOs confer legitimacy on policy decisions and thus increase the prospect that such policies will be implemented (see Breitmeier 2005; Zürn 2004). Although our study did not explicitly examine the relationship between NGO influence and the problem-solving performance of international environmental agreements, we did find that NGO influence was highest when the political stakes of the negotiations were lowest. In other words, NGO information appears the have the greatest effect when the negotiations involve limited commitments for behavioral change, so we must question the claim that such influence necessarily results in more effective problem-solving.

Our cases confirm also that NGOs provide information in a highly political context where there are debates about how to interpret expert knowledge and/or where information is seen to threaten certain interests. Translating information provided by NGOs into specific policies is rarely a straightforward matter.[9] In addition the link between NGO participation/influence in international environmental negotiations and policy implementation requires further investigation. Hochstetler (2002) argues that domestic implementation depends not only on what happened during the negotiation process but also on whether a state has accepted the relevant international norms and its domestic capacities (see also Risse, Ropp, and Sikkink 1999). This again raises the idea of multilevel governance and suggests that NGOs wishing to shape environmental outcomes may need to look beyond multilateral negotiations and work in multiple spheres and tiers of governance simultaneously.

Democratizing Global Governance

Scholars and policy-makers argue that NGO participation in multilateral negotiations on the environment and sustainable development democratizes global governance (e.g., Raustiala 1997; Willetts 1996b; Princen 1994). In some respects our cases show that international environmental negotiations have become more democratic over time in the sense that states cannot legitimately exclude NGOs from decision-making processes. In all of our cases, states provided some space for NGOs to voice their views. There was never a question of whether NGOs would be permitted to participate; rather states and NGOs debated over the specific details of how NGO diplomats would participate. Nevertheless, we also saw many instances where states tried to resist NGO demands for participation by placing limits on the opportunities for NGO participation in order to maintain control over the negotiation process (see also Friedman, Hochstetler, and Clark 2005).

However, democratizing global governance involves more than increasing the number of participants involved in multilateral decision-making processes. Our studies highlight the need to take seriously the issues of NGO accountability and representation (Jordan and Van Tuijl 2000; Held 1999; Chartier and Deleage 1998; Pasha and Blaney 1998). By their political nature, NGOs, like states, have well-defined interests, and they act strategically to pursue those interests. When evaluating the effect of NGOs on democracy, it is important to ask who these groups represent and to what extent they are accountable to their constituents and/or one another. For example, what are the implications for representation when NGOs receive significant funding from state-based institutions or if members of the South are systematically disadvantaged (Duwe 2001; Yamin 2001; Kellow 2000; Chartier and Deleage 1998)? Jordan and Van Tuijl (2000: 2061) poignantly ask how NGOs can democratize institutions of global governance if they "reflect as much inequality as they are trying to undo?" Suggestions that some NGOs employ questionable tactics, such as manipulating scientific findings, raise further questions about accountability (Skodvin and Andresen 2003; Harper 2001; Jordan 2001; Tesh 2000).

Diplomacy and World Politics
Our project clearly demonstrates the changing nature of diplomacy in world politics. Multilateral negotiation processes to address global environmental challenges cannot be understood in terms of inter-state diplomacy. These processes involve myriad actors representing a diversity of interests. Some interests are territorially defined (e.g., national sovereignty), while others are defined in terms of common moral, scientific, and/or economic concerns. The line between official and unofficial forms of diplomacy is increasingly becoming blurred as NGOs directly engage in one of the most traditional diplomatic activities: the negotiation of multilateral agreements. In such settings NGOs perform many of the same functions as state delegates: they represent the interests of their constituencies, they engage in information exchange, they negotiate, and they provide policy advice (Jönsson 2002). Time and again our contributors noted the professionalism of NGO representatives serving in this capacity.

The changing nature of diplomacy in international environmental negotiations reflects the increasing complexity of world politics. States are recognizing that they can no longer address all the challenges facing the international community on their own, so they are drawing on the expertise and resources that NGOs have to offer. In addition NGOs are demanding a more central position in international decision-making processes, challenging notions of state sovereignty. Together, these trends have resulted in a realignment of state–society relations in international politics. Scholars of international environmental negotiations as well as practitioners face the daunting task of trying to understand what has become a multi-level game involving diplomacy between and within states, between and within NGOs, and between NGOs and states at multiple levels of social organization.

Notes

1. We acknowledge that many NGOs are more interested in networking with other NGOs at negotiating sessions than engaging in NGO diplomacy. However, as we discussed in chapter 1, our primary interest is in analyzing the influence of those non-state actors that set out to shape negotiating processes and outcomes.

2. George and Bennett (2005: 71) argue that structure and focus are easier to achieve when a single investigator plans and conducts the case studies: "Properly coordinating the work of case writers in a collaborative study can be a challenging task for the chief investigator, particularly when the contributors are well-established scholars with views of their own regarding the significance of the case they are preparing." We wholeheartedly agree with this observation.

3. We are grateful to David Humphreys for raising this issue at the Stockholm Workshop. See Corell and Betsill (2003).

4. We did not draw on the new management procedure case very heavily in our comparative analysis because it did not focus directly on the influence of scientists' in the actual decision to adopt a new management procedure. Rather, Andresen and Skodvin have provided a more a general discussion of scientists' influence during the 1970s, both before and after the decision.

5. For details of these assessments, please contact the author.

6. Scholars commonly distinguish among three distinct (but overlapping) phases of international policy making: agenda-setting/pre-negotiation, policy formulation/negotiation, and implementation (see Newell 2000; Young 1997). Our focus in this volume has been on the policy formulation/negotiation stage.

7. Newell (2000) and Rowlands (2000) argue that industry groups often influence environmental politics via the trade regime.

8. Thanks to Kathy Hochstetler for calling this to our attention.

9. For example, see the vast literature on the science-policy interface: Dimitrov (2003), Harrison and Bryner (2003), Andresen et al. (2000), and Jasanoff and Wynne (1998).

References

The Access Initiative. 2005. Home Page. Available at ⟨http://www .accessinitiative.org/⟩ (accessed 1 August 2005).

Ad Hoc Group on the Berlin Mandate. 1996. *Report of the Ad Hoc Group on the Berlin Mandate on the Work of its Fourth Session, Geneva, 11–16 July 1996* (7 October 1996). UN Document FCCC/AGBM/1996/8.

Ad Hoc Group on the Berlin Mandate. 1997a. *Compilation of Responses from Parties on Issues Related to Sinks* (29 November). Note by the Secretariat, UN Document FCCC/AGBM/1997/INF.2.

Ad Hoc Group on the Berlin Mandate. 1997b. *Framework Compilation of Proposals from Parties for the Elements of a Protocol or Another Legal Instrument* (31 January 1997). UN Document FCCC/AGBM/1997/2.

Alter, K. J., and S. Meunier. 2006. Nested and overlapping regimes in the transatlantic banana trade dispute. *Journal of European Public Policy* 13 (3): 362–82.

American Lands Alliance. 2000. The free trade area of the Americas: Hemispheric forest threat. Available at ⟨http://www.americanlands.org/IMF/free_trade.htm⟩ (accessed 29 July 2003).

Andersen, R. 2002. The time dimension in international regime interplay. *Global Environmental Politics* 2 (3): 98–117.

Andresen, S. 1998. The making and implementation of whaling policies: Does participation make a difference? In *The Implementation and Effectiveness of International Commitments: Theory and Practice*, edited by D. G. Victor, K. Raustiala, and E. B. Skolnikoff. Cambridge: MIT Press, pp. 431–74.

Andresen, S. 2000. The whaling regime. In *Science and Politics in International Environmental Regimes: Between Integrity and Involvement*, edited by S. Andresen, T. Skodvin, A. Underdal, and J. Wettestad. Manchester: Manchester University Press, pp. 35–70.

Andresen, S. 2001. The whaling regime: 'Good' policy but 'bad' institutions? In *Towards a Sustainable Whaling Regime?* edited by R. L. Friedheim. Seattle: University of Washington Press, 235–65.

Andresen, S. 2002. The International Whaling Commission: More failure than success? In *Environmental Regime Effectiveness*, edited by E. Miles, A. Underdal, S. Andresen, J. Wettestad, J. B. Skjærseth, and E. Carlin. Cambridge: MIT Press, 379–404.

Andresen, S. 2004. Norwegian whaling policy: Peace at home, war on the international scene. In *International Regimes and Norway's Environmental Policy: Crossfire and Coherence*, edited by J. B. Skjærseth. Aldershot: Ashgate Publishing, pp. 41–64.

Andresen, S., T. Skodvin, A. Underdal, and J. Wettestad, eds. 2000. *Science and Politics in International Environmental Regimes*. Manchester, UK: Manchester University Press.

Ansell, C., R. Maxwell, and D. Sicurelli. 2006. Protesting food: NGOs and political mobilization in Europe. In *What's the Beef? The Contested Governance of European Food Safety*, edited by C. Ansell and D. Vogel. Cambridge: MIT Press, 97–122.

Arts, B. 1998. *The Political Influence of Global NGOs: Case Studies on the Climate and Biodiversity Conventions*. Utrecht: International Books.

Arts, B. 2001. Impact of ENGOs on international conventions. In *Non-state Actors in International Relations*, edited by B. Arts, M. Noortmann, and B. Reinalda. Aldershot, UK: Ashgate Publishing, pp. 195–210.

Arts, Bas, and S. Mack. 2003. Environmental NGOs and the biosafety protocol: A case study on political influence. *European Environment* 13: 19–33.

Auer, M. 1998. Colleagues or combatants? Experts as environmental diplomats. *International Negotiation* 3 (2): 267–87.

Aviel, J. F. 2005. NGOs and international affairs. In *Multilateral Diplomacy and the United Nations Today*, 2nd ed., edited by J. P. Muldoon Jr., J. F. Aviel, R. Reitano, and E. Sullivan. Boulder, CO: Westview Press, pp. 159–72.

Axelrod, R. 1984. *The Evolution of Cooperation*. New York: Basic Books.

Bail, C., J. P. Decaestecker, and M. Jorgensen. 2002. European Union. In *The Cartagena Protocol on Biosafety: Reconciling Trade in Biotechnology with Environment and Development?* edited by C. Bail, R. Falkner, and H. Marquard. London: Earthscan, pp. 166–85.

Bailey, J. 2006. Arrested development: The prohibition of commercial whaling as a case of failed norm change. Paper presented at the 47th Annual Convention of the International Studies Association, San Diego, CA, 21–26 March.

Benedick, R. E. 1991. *Ozone Diplomacy: New Directions in Safeguarding the Planet*. Cambridge: Harvard University Press.

Bernauer, T., and E. Meins. 2003. Technological revolution meets policy and the market: Explaining cross-national differences in agricultural biotechnology regulation. *European Journal of Political Research* 42 (5): 643–84.

Bernow, S., W. Dougherty, M. Duckworth, S. Kartha, M. Lazarus, and M. Ruth. 1997. *Policies and Measures to Reduce CO_2 Emissions in the United States: An*

Analysis of Options for 2005 and 2010. Boston: Tellus Institute for Resource and Environmental Strategies and Stockholm Environment Institute.

Bernstein, J., P. Chasek, and L. J. Goree. 1993a. A brief history of the INCD. *Earth Negotiations Bulletin* 4 (2).

Bernstein, J., P. Chasek, and I. J. Goree. 1993b. A summary of the Proceedings of the Organizational Session of the INC for the Elaboration of an International Convention to Combat Desertification. *Earth Negotiations Bulletin* 4 (1).

Bernstein, J., P. Chasek, L. J. Goree, and W. Mwangi. 1994. Summary of the fifth session of the INC for the Elaboration of an International Convention to Combat Desertification, 6–17 June 1994. *Earth Negotiations Bulletin* 4 (55).

Bernstein, S. 2001. *The Compromise of Liberal Environmentalism.* New York: Columbia University Press.

Betsill, M. M. 2000. *Greens in the Greenhouse: Environmental NGOs, Norms and the Politics of Global Climate Change.* PhD thesis. Department of Political Science, University of Colorado, Boulder.

Betsill, M. M. 2002. Environmental NGOs meet the sovereign state: The Kyoto Protocol negotiations on global climate change. *Colorado Journal of International Environmental Law and Policy* 13 (1): 49–64.

Betsill, M. M. 2004. Global climate change policy: Making progress or spinning wheels? In *The Global Environment: Institutions, Law and Policy,* 2nd ed., edited by R. S. Axelrod, D. Downie, and N. J. Vig. Washington, DC: CQ Press, pp. 103–24.

Betsill, M. M. 2006. Transnational Actors in International Environmental Politics. In *Palgrave Advances in the Study of International Environmental Politics,* edited by M. M. Betsill, K. Hochstetler, and D. Stevis. Basingstoke: Palgrave Macmillan, pp. 172–202.

Betsill, M. M., and H. Bulkeley. 2006. Cities and the multilevel governance of global climate change. *Global Governance* 12 (2): 141–59.

Betsill, M. M., and E. Corell. 2001. NGO Influence in international environmental negotiations: A framework for analysis. *Global Environmental Politics* 1 (4): 65–85.

Bettelli, P., C. Carpenter, D. Davenport, and P. Doran. 1997. Report of the Third Conference of the Parties to the United Nations Framework Convention on Climate Change, 1–11 December 1997. *Earth Negotiations Bulletin* 12 (76).

Biermann, F. 2000. The case for a World Environment Organization. *Environment* 42 (9): 22–31.

Biersteker, T. J. 1995. Constructing historical counterfactuals to assess the consequences of international regimes. In *Regime Theory and International Relations,* edited by V. Rittberger. Oxford: Clarendon Press, pp. 315–38.

Biliouri, D. 1999. Environmental NGOs in Brussels: How powerful are their lobbying activities? *Environmental Politics* 8: 173–82.

Blaikie, P., and H. Brookfield. 1987. *Land Degradation and Society*. London: Routledge.

BNA. 1997. Business, Labor, Agriculture Coalition Sponsors Ad Campaign against Climate Treaty. *BNA International Environment Daily*, 10 September.

Boehmer-Christiansen, S. 1994. Global climate protection policy: The Limits of scientific advice, Part 1 and Part 2. *Global Environmental Change* 4 (2–3): 140–59, 185–200.

Boulton, L., and B. Hutton. 1997. Kyoto climate change talks agree treaty at last minute. *Financial Times (London)*, 11 December.

Breitmeier, H. 2005. *The Legitimacy of International Regimes*. Aldershot: Ashgate.

Brown, P., and J. Leggett. 1997. Environment: Only way is up. *The Guardian (London)*, 17 December.

Bullock, D., M. Desquilbet, and E. Nitsi. 2000. *The Economics of Non-GMO Segregation and Identity Preservation*. Urbana, IL: University of Illinois Press.

Burgiel, S. 2002. Negotiating the trade-environment frontier: Biosafety and intellectual property rights in international policy-making. PhD dissertation, American University, Washington, DC.

Carpenter, C. 2001. Businesses, Green groups and the media: The role of non-governmental organizations in the climate change debate. *International Affairs* 77 (2): 313–28.

Charnovitz, S. 1997. Two centuries of participation: NGOs and international governance. *Michigan Journal of International Law* 18 (2): 183–286.

Chartier, D., and J.-P. Deleage. 1998. The international environmental NGOs: From the revolutionary alternatives to the pragmatism of reform. *Environmental Politics* 7: 26–41.

Chasek, P. S. 2001. *Earth Negotiations: Analyzing Thirty Years of Environmental Diplomacy*. New York: United Nations University Press.

Chasek, P., E. Corell, L. J. Goree, and W. Mwangi. 1995. Summary of the sixth session of the INC for the Elaboration of an International Convention to Combat Desertification, 9–18 January 1995. *Earth Negotiations Bulletin* 4 (65).

Chasek, P., L. J. Goree, and W. Mwangi. 1993. Summary of the first session of the INC for the Elaboration of an International Convention to Combat Desertification, 24 May–3 June 1993. *Earth Negotiations Bulletin* 4 (11).

Chatterjee, P., and M. Finger. 1994. *The Earth Brokers: Power, Politics and World Development*. London: Routledge.

Clark, A. M., E. J. Friedman, and K. Hochstetler. 1998. The sovereign limits of global civil society: A comparison of NGO participation in UN world conferences on the environment, human rights and women. *World Politics* 51: 1–35.

Climate Action Network. 2003. *About Climate Action Network*. Available at ⟨http://www.climatenetwork.org/⟩ (accessed 3 June 2003).

Close, D. 1998. Environmental NGOs in Greece: The Acheloos Campaign as a case study. *Environmental Politics* 7: 55–77.

Conference of the Parties. 1995. The Berlin Mandate: Review of the adequacy of Article 4, paragraph 2(a) and (b), of the Convention, including proposals related to a protocol and decisions on follow-up. *Report of the Conference of the Parties on its First Session*, Berlin, 28 March–7 April 1995. *Addendum. Part Two: Actions Taken by the Conference of the Parties at its First Session*, 6 June 1995. UN Document FCCC/CP/1995/7/Add.1.

Corell, E. 1998. North-South financial tensions: Desertification after UNGASS. *Environmental Politics* 7 (1): 222–26.

Corell, E. 1999a. *The Negotiable Desert: Expert Knowledge in the Negotiations of the Convention to Combat Desertification*. PhD thesis. Linköping Studies in Arts and Science 191. Linköping University, Sweden.

Corell, E. 1999b. Non-state actor influence in the negotiations of the Convention to Combat Desertification. *International Negotiation* 4 (2): 197–223.

Corell, E., and M. M. Betsill. 2001. A comparative look at NGO influence in international environmental negotiations: Desertification and climate change. *Global Environmental Politics* 1 (4): 86–107.

Corell, E., W. Mwangi, and S. Wise. 1996. Summary of the eighth session of the INC for the Convention to Combat Desertification. *Earth Negotiations Bulletin* 4 (86).

Corell, E., W. Mwangi, T. Prather, and L. Wagner. 1997a. Highlights of CCD COP-1 Thursday, 9 October 1997. *Earth Negotiations Bulletin* 4 (115).

Corell, E., W. Mwangi, T. Prather, and L. Wagner. 1997b. Summary of the First Conference of the Parties to the Convention to Combat Desertification, 29 September–10 October 1997. *Earth Negotiations Bulletin* 4 (116).

Cosbey, A., and S. Burgiel. 2000. The Cartagena Protocol on Biosafety: An analysis of results. *IISD Briefing Note*. Winnipeg: International Institute for Sustainable Development.

Cox, R. W., and Harold K. Jacobson. 1973. *The Anatomy of Influence: Decision Making in International Organization*. New Haven: Yale University Press.

Cushman Jr., J. H., and D. E. Sanger. 1997. Global warming, no simple fight: The forces that shaped the Clinton plan. *The New York Times*, 1 December, p. F3.

Dahl, R. 1957. The concept of power. *Behavioral Science* 2: 201–15.

de Greef, W. 2000. Regulatory conflicts and trade. *NYU Environmental Law Review* (8): 579–84.

Depledge, J. 2000. Tracing the origins of the Kyoto Protocol: An article-by-article textual history (25 November). Technical Paper FCCC/TP/2000/2.

DeSombre, E. R. 2000. *Domestic Sources of International Environmental Policy: Industry, Environmentalists, and U.S. Power*. Cambridge: MIT Press.

DeSombre, E. 2001. Distorting global governance: Membership, voting, and the IWC. In *Towards a Sustainable Whaling Regime*, edited by R. L. Friedheim. Seattle: University of Washington Press, pp. 183–200.

Dimitrov, R. S. 2003. Knowledge, power, and interests in environmental regime formation. *International Studies Quarterly* 47: 123–50.

Dodds, F. 2001. From the corridors of power to the global negotiating table: The NGO steering committee of the Commission on Sustainable Development. In *Global Citizen Action*, edited by M. Edwards and J. Gaventa. Boulder, CO: Lynne Rienner, pp. 203–13.

Dodds, F. 2002. The Context: Multi-stakeholder Processes and Global Governance. In *Multi-Stakeholder Processes for Governance and Sustainability: Beyond Deadlock and Conflict*, edited by M. Hemmati. London: Earthscan, pp. 26–38.

Donovan, R. Z. 1996. Role of NGOs. In *Certification of Forest Products: Issues and Perspectives*, edited by V. M. Viana, J. Ervin, R. Z. Donovan, C. Elliott, and H. Gholz. Washington, DC: Island Press, pp. 93–110.

Duwe, M. 2001. The climate action network: Global civil society at work? *Reciel* 10 (2): 1–14.

Earth Negotiations Bulletin. 1992. Convention to Combat Desertification (A/C.2/47/L.46), 3 (3).

European Eco Forum. *Public Participation Campaign* 2005. Available at ⟨http:// www.participate.org/work_programme/pp_campaign.htm⟩ (accessed 1 August 2005).

Falkner, R. 2003. Private environmental governance and international relations: Exploring the links. *Global Environmental Politics* 3 (2): 72–87.

Fearon, J. D. 1998. Bargaining, enforcement and international cooperation. *International Organization* 52: 269–305.

Federal Register Notice. 1999. *Federal Register Notice on WTO Forest Products Agreement, 14 July 1999.* Available at ⟨http://forests.org/archive/general/federeg .htm⟩ (accessed 29 July 2003).

FERN. 1999. *NGO Statement of Opposition to the Proposed Liberalization of the Forest Products Sector.* Available at ⟨http://www.fern.org/pubs/archive/ liberal2.htm⟩ (accessed 29 July 2003).

Frank, D. J., A. Hironaka, J. W. Meyer, E. Schofer, and N. B. Tuma. 1999. The Rationalization and Organization of Nature in World Culture. In *Constructing World Culture: International Nongovernmental Organizations since 1875*, edited by J. Boli and G. Thomas. Stanford: Stanford University Press, pp. 81–99.

Freestone, D., and E. Hey, eds. 1995. *The Precautionary Principle and International Law: The Challenge of Implementation.* Boston: Kluwer Law International.

Friedheim, R. L. 1996. Moderation in pursuit of justice: Explaining Japan's failure in the international whaling negotiations. *Ocean Development and International Law* 27: 349–78.

Friedheim, R. L., ed. 2001. *Toward a Sustainable Whaling Regime*. Seattle: University of Washington Press.

Friedman, E. J., K. Hochstetler, and A. M. Clark. 2005. *Sovereignty, Democracy and Global Civil Society: State–Society Relations at UN World Conferences*. Ithaca, NY: SUNY Press.

Gale, F. P. 1998a. Constructing global civil society actors: An anatomy of the environmental coalition contesting the tropical timber trade regime. *Global Society* 12 (3): 343–61.

Gale, F. P. 1998b. *The Tropical Timber Trade Regime*. London: Macmillan.

Gale, L. 2002. Greenpeace International. In *The Cartagena Protocol on Biosafety: Reconciling Trade in Biotechnology with Environment and Development?* edited by C. Bail, R. Falkner, and H. Marquard. London: Earthscan, pp. 251–62.

Gambell, R. 1995. Management of Whaling in Coastal Communities. In *Whales, Seals, Fish and Man*, edited by A. S. Blix, L. Walløe, and Ø. Ulltang. Amsterdam: Elsevier Science, pp. 699–708.

Gamson, W. A., and D. S. Meyer. 1996. Framing political opportunity. In *Comparative Perspectives on Social Movements: Political Opportunities, Mobilizing Structures, and Cultural Framings*, edited by D. McAdam, J. J. McCarthy, and M. N. Zald. Cambridge: Cambridge University Press, pp. 275–90.

Gemmill, B., and Bamidele-Izu. 2002. The role of NGOs and civil society in global environmental governance. In *Global Environmental Governance: Options and Opportunities*, edited by D. C. Esty and M. H. Ivanova. New Haven: Yale School of Forestry and Environmental Studies, pp. 77–100.

George, A. L., and A. Bennett. 2005. *Case Studies and Theory Development in the Social Sciences*. Cambridge: MIT Press.

Gereffi, G., R. Garcia-Johnson, and E. Sasser. 2001. The NGO-industrial complex. *Foreign Policy* (July–August): 56–65.

Giugni, M. 1999. How social movements matter: Past research, present problems, future developments. In *How Social Movements Matter*, edited by M. Giugni, D. McAdam and C. Tilly. Minneapolis: University of Minnesota Press, pp. xiii–xxxiii.

Glantz, M. H., and N. Orlovsky. 1983. Desertification: A review of the concept. *Desertification Control Bulletin* 9: 15–22.

Global Forest Policy Project. 1995. The Open-Ended Inter-Governmental Panel on Forests—Proposed Terms of Reference, April 11, 1995. Statement to 1995 meeting of UN Commission on Sustainable Development.

Global Industry Coalition. February 1999. Biosafety Negotiations: Industry Expresses Caution and Optimism. Cartagena.

Glück, P., R. Tarasofsky, N. Byron, and I. Tikkanen. 1997. *Options for Strengthening the International Legal Regime for Forests*. Joensuu, Finland: European Forest Institute.

Gordenker, L., and T. G. Weiss, eds. 1996. *NGOs, the UN and Global Governance*. Boulder, CO: Lynne Reinner.

Gore, A. 1997. Remarks by Vice President Al Gore, United Nations Committee on Climate Change, Conference of the Parties, 8 December 1997, Kyoto, Japan.

Griffiths, T. 2001. *Consolidating the Gains: Indigenous Peoples; Rights and Forest Policy Making at the United Nations*. A Forest Peoples Programme Briefing Paper. Available at ⟨http://forestpeoples.gn.apc.org/Briefings/UNFF/briefing_unff_&_iprights_dec01⟩ (accessed 21 July 2003).

Grolin, J. 1996. Aarhus: Negotiating around a biosafety protocol. *Biodiversity Bulletin* 1 (2): 5–6.

Grubb, M., C. Vrulijk, and D. Brack. 1999. *The Kyoto Protocol: A Guide and Assessment*. London: Royal Institute of International Affairs.

Gulbrandsen, L. H., and S. Andresen. 2004. NGO Influence in the implementation of the Kyoto Protocol: Compliance, flexibility mechanisms, and sinks. *Global Environmental Politics* 4 (4): 54–75.

Gutman, P. 2003. What did the WSSD accomplish? *Environment* 45 (2): 20–26.

Haas, P. M. 1990. *Saving the Mediterranean: The Politics of International Environmental Cooperation*. New York: Columbia University Press.

Haas, P. M. 1992. Introduction: Epistemic communities and international policy coordination. *International Organization* 46 (1): 1–35.

Hare, F. K. 1993. *Climate Variations, Drought and Desertification*. Geneva: WMO.

Harper, C. 2001. Do the facts matter? NGOs, research, and international advocacy. In *Global Citizen Action*, edited by M. Edwards and J. Gaventa. Boulder, CO: Lynne Rienner, pp. 247–58.

Harrison, N. E., and G. Bryner, eds. 2003. *Science and Politics in the International Environment*. Lanham, MD: Rowman and Littlefield.

Held, D. 1999. The Transformation of Political Community: Rethinking Democracy in the Context of Globalization. In *Democracy's Edges*, edited by I. Shapiro and C. Hacker-Cordón. Cambridge: Cambridge University Press, pp. 84–111.

Helldén, U. 1991. Desertification—Time for an assessment? *Ambio* 20 (8): 372–83.

Hemmati, M. 2002. *Multi-stakeholder Processes for Governance and Sustainability: Beyond Deadlock and Conflict*. London: Earthscan.

Hochstetler, K. 2002. After the boomerang: Environmental movements and politics in the La Plata River Basin. *Global Environmental Politics* 2 (4): 35–57.

Holsti, K. J. 1988. *International Politics: A Framework for Analysis*. Toronto: Prentice-Hall International.

Hooghe, L., and G. Marks. 2003. Unraveling the central state, but how? Types of multi-level governance. *American Political Science Review* 97 (2): 233–43.

Hulme, D., and M. Edwards, eds. 1997. *NGOs, States and Donors: Too Close for Comfort?* Basingstoke: Macmillan Press.

Humphreys, D. 1996a. *Forest Politics: The Evolution of International Cooperation.* London: Earthscan.

Humphreys, D. 1996b. Hegemonic ideology and the international tropical timber organization. In *The Environment and International Relations*, edited by J. Vogler and M. F. Imber. London: Routledge, pp. 215–33.

Humphreys, D. 1996c. Regime theory and non-governmental organizations: The case of forest conservation. *Journal of Commonwealth and Comparative Politics* 34 (1): 90–115.

Humphreys, D. 2001. Forest negotiations at the United Nations: Explaining cooperation and discord. *Forest Policy and Economics* 3 (3–4): 125–35.

Humphreys, D. 2003. Life protective or carcinogenic challenge: Global forests governance under advanced capitalism. *Global Environmental Politics* 3 (2): 40–55.

Humphreys, D. 2004. Redefining the issues: NGO influence on international forestry negotiations. *Global Environmental Politics* 4 (2): 51–74.

Humphreys, D. 2006. *Logjam: Deforestation and the Crisis of Global Governance.* London: Earthscan.

INCD Document A/AC.241/CRP.2, 31 May 1993. Conference Room Paper: Statement by Mr. Hama Arba Diallo, Executive Secretary of INCD, at the opening of the discussion on item 4 on Monday, 31 May 1993.

INCD Secretariat. 1993. Report of the first session of the International Panel of Experts on Desertification, 22–25 February 1993, March.

International Centre for Trade and Sustainable Development. 1999. Accreditation schemes and other arrangements for public participation in international fora: A contribution to the Debate on WTO and transparency. Geneva: International Centre for Trade and Sustainable Development.

International Institute for Sustainable Development (IISD). 2001. UNFF Final. *Earth Negotiations Bulletin* 13 (83). Available at ⟨http://www/iisd.ca/forestry/unff/unff1/⟩ (accessed 25 June 2001).

International NGO Conference on Desertification. 1993. Proposals to the Intergovernmental Negotiating Committee for a Convention to Combat Desertification, 16–20 August, Bamako, Mali.

International Tropical Timber Organization (ITTO). 1989a. ITTO document PCM, PCF, PCI(V)/1. Pre-project Proposal: Labelling Systems for the Promotion of Sustainably-Produced Tropical Timber, 15 August.

International Tropical Timber Organization (ITTO). 1989b. ITTO document PCM(V)/D.1. Report to the International Tropical Timber Council, fifth session of the Permanent Committee on Economic Information and Market Intelligence, 3 November.

IPCC. 1996. *Climate Change 1995: The Science of Climate Change, Summary for Policymakers.* Cambridge: Cambridge University Press. Available at ⟨http://www.ipcc.ch/pub/sarsum1.htm#four⟩.

Jasanoff, S., and B. Wynne. 1998. Science and decision-making. In *Human Choice and Climate Change: The Societal Framework*, edited by S. Rayner and E. Malone. Columbus, OH: Batelle Press, pp. 1–87.

Jaura, R. 1997. Environment: U.S. Greenhouse Gas Proposal Disappoints Many. *Inter Press Service*, 22 October.

Jönsson, C. 2002. Diplomacy, Bargaining and Negotiation. In *Handbook of International Relations*, edited by W. Carlsnaes, T. Risse, and B. A. Simmons. Thousand Oaks, CA: Sage, pp. 212–34.

Jordan, G. 2001. *Shell, Greenpeace and the Brent Spar.* Basingstoke: Palgrave.

Jordan, L., and P. Van Tuijl. 2000. Political responsibility in transnational NGO advocacy. *World Development* 28: 2051–65.

Kakabadse, Y. N., with S. Burns. 1994. *Movers and Shapers: NGOs in International Affairs* (May). Washington, DC: World Resources Institute.

Kalland, A. 1993. Management by totemization: Whale symbolism and the anti-whaling campaign. *Arctic* 46 (2): 124–33.

Kassas, M. 1995. Negotiations for the International Convention to Combat Desertification (1993–1994). *International Environmental Affairs* 7 (2): 176–86.

Kay, T. 2005. Labor transnationalism and global governance: The impact of NAFTA on transnational labor relations in North America. *American Journal of Sociology* 111 (3): 715–56.

Keck, M. E., and K. Sikkink. 1998. *Activists Beyond Borders: Advocacy Networks in International Politics.* Ithaca, NY: Cornell University Press.

Kellow, A. 2000. Norms, interests and environmental NGOs: The limits of cosmopolitanism. *Environmental Politics* 9 (3): 1–22.

Khagram, S., J. V. Riker, and K. Sikkink. 2002. From Santiago to Seattle: Transnational advocacy groups restructuring world politics. In *Restructuring World Politics: Transnational Social Movements, Networks and Norms*, edited by S. Khagrm, J. V. Riker, and K. Sikkink. Minneapolis: University of Minnesota Press, pp. 3–23.

King, G., R. O. Keohane, and S. Verba. 1994. *Designing Social Inquiry: Scientific Inference in Qualitative Research.* Princeton: Princeton University Press.

Knight, D. 1997. Environment–U.S.: Million Spent in Fight against Climate Treaty. *Inter Press Service*, 12 September 1997.

Knocke, D. 1990. *Political Networks: The Structural Perspective.* Cambridge: Cambridge University Press.

Kremenyuk, V. A., and W. Lang. 1993. The political, diplomatic, and legal background. In *International Environmental Negotiation*, edited by G. Sjöstedt. Newbury Park: Sage, pp. 3–16.

Krueger, J. 1999. Trade restrictions and the Montreal Protocol. In *Environmental Issues in North–South Trade Negotiations*, edited by D. Tussie. New York: Macmillan.

Levy, D. L., and P. Newell. 2000. Oceans apart? Business responses to global environmental issues in Europe and the United States. *Environment* 42 (9): 8–20.

Liberatore, A. 1995. The social construction of environmental problems. In *Environmental Policy in an International Context: Perspectives on Environmental Problems*, edited by P. Glasbergen and A. Blowers. London: Arnold, pp. 59–83.

Madeley, J. 2000. *Hungry for Trade: How the Poor Pay for Free Trade*. London: Zed.

Magalhães, J. C. De. 1988. *The Pure Concept of Diplomacy*. New York: Greenwood Press.

Mainguet, M. 1994. *Desertification: Natural Background and Human Mismanagement*. Berlin: Springer-Verlag.

Mann, M. 1999. EU loath to use national law on national GMO bans. *Reuters Financial Report*, 10 December.

Marton-Lefèvre, J. 1994. The role of the scientific community in the preparation of and follow-up to UNCED. In *Negotiating International Regimes: Lessons Learned from the United Nations Conference on Environment and Development*, edited by B. I. Spector, G. Sjöstedt, and I. W. Zartman. London: Graham and Trotman, 171–80.

Mathews, J. T. 1997. Power shift. *Foreign Affairs* 76 (1): 50–66.

McAdam, D. 1996. Conceptual origins, current problems, future directions. In *Comparative Perspectives on Social Movements: Political Opportunities, Mobilizing Structures, and Cultural Framings*, edited by D. McAdam, J. J. McCarthy, and M. N. Zald. Cambridge: Cambridge University Press, pp. 23–40.

McHugh, J. L. 1974. The role and history of the International Whaling Commission. In *The Whale Problem*, edited by W. E. Schevill. Cambridge: Harvard University Press, pp. 305–35.

Meyer, A. 1990. The campaign continues. *Geographical Magazine* (February): 14–15.

Middleton, N. J., and D. S. G. Thomas, eds. 1997. *World Atlas of Desertification*, 2nd ed. London: Arnold Publishers and United Nations Environment Program.

Miles, M. B., and A. M. Huberman. 1994. *Qualitative Data Analysis*, 2nd ed. Thousand Oaks, CA: Sage.

Mintzer, I. M., and J. A. Leonard, eds. 1994. *Negotiating Climate Change: The Inside Story of the Rio Convention*. Cambridge: Cambridge University Press.

Mitchell, R. B. 2002. A quantitative approach to evaluating international environmental regimes. *Global Environmental Politics* 2 (4): 58–83.

Mitchell, R. B., and T. Bernauer. 1998. Empirical research on international environmental policy: Designing qualitative case studies. *Journal of Environment and Development* 7 (1): 4–31.

Morphet, S. 1996. NGOs and the environment. In *The Conscience of the World: The Influence of Non-governmental Organisations in the U.N.*, edited by P. Willetts. Washington, DC: Brookings Institute, pp. 116–46.

National Council for International Visitors. 2006. *Welcome.* Available at ⟨http://www.nciv.org⟩ (accessed 29 August 2006).

Newell, P. 2000. *Climate for Change: Non-state Actors and the Global Politics of the Greenhouse.* Cambridge: Cambridge University Press.

NGO Statement at PrepCom 4. 1992. NGO Recommendations for Textual Amendments to Forest-Related Documents. *Forest Principles* (CRP.14/Rev.2) and *Agenda 21* (PC/100/ADD.16), PrepcomIV, 16 March.

Nijar, G. S. 1997. Liability and compensation in a biosafety protocol. TWN Paper 4. Penang.

Nitze, W. A. 1994. A failure of presidential leadership. In *Negotiating Climate Change: The Inside Story of the Rio Convention*, edited by I. M. Mintzer and J. A. Leonard. Cambridge: Cambridge University Press, pp. 187–200.

Nye Jr., J. S. 1990. *Bound to Lead: The Changing Nature of American Power.* New York: Basic Books.

Oberthür, S., and H. E. Ott. 1999. *The Kyoto Protocol: International Climate Policy for the 21st Century.* New York: Springer.

Oberthür, S., M. Buck, S. Müller, S. Pfahl, R. G. Tarasofsky, J. Werksman, and A. Palmer. 2002. *Participation of Non-governmental Organisations in International Environmental Governance: Legal Basis and Practical Experience.* Berlin: Ecologic-Institute for International and European Environmental Policy.

Odingo, R. S., ed. 1990. *Desertification Revisited: Proceedings of an Ad hoc Consultative Meeting on the Assessment of Desertification.* Nairobi: UNEP-DC/PAC.

Otinda, P., B. Ibrahima, and R. Sales Jr. 1997. South Counters US Proposal. *ECO,* 4 March.

Parson, E. 2003. *Protecting the Ozone Layer: Science and Strategy.* Oxford: Oxford University Press.

Pasha, M. K., and D. L. Blaney. 1998. Elusive paradise: The promise and peril of global civil society. *Alternatives* 23: 417–50.

Paterson, M. 1996. *Global Warming and Global Politics.* London: Routledge.

Peterson, M. J. 1992. Whalers, cetologists, environmentalists, and the international management of whaling. *International Organization* 46 (1): 147–86.

Pomerance, R. 2000. The biosafety protocol: Cartagena and beyond. *NYU Environmental Law Review* (8): 614–21.

Poore, D. 1989. *No Timber without Trees: Sustainability in the Tropical Forest.* London: Earthscan.

Poore, D. 2003. *Changing Landscapes: The Development of the International Tropical Timber Organization and Its Influence on Tropical Forest Management.* London: Earthscan.

Potter, D., ed. 1996. *NGOs and Environmental Policies: Asia and Africa.* London: Frank Cass.

Princen, T., and M. Finger, eds. 1994. *Environmental NGOs in World Politics: Linking the Local and the Global.* London: Routledge.

Raustiala, K. 1997. States, NGOs and international environmental institutions. *International Studies Quarterly* 41: 719–40.

Raustiala, K. 2001. Nonstate actors in the global climate regime. In *International Relations and Global Climate Change*, edited by U. Luterbacher and D. F. Sprinz. Cambridge: MIT Press, pp. 95–118.

Risse, T., S. C. Ropp, and K. Sikkink, eds. 1999. *The Power of Human Rights: International Norms and Domestic Change.* Cambridge: Cambridge University Press.

Rosendal, G. K. 2001a. Impact of overlapping international regimes: The case of biodiversity. *Global Goverance* 7 (2): 95–117.

Rosendal, G. K. 2001b. Overlapping international regimes: The case of the Intergovernmental Forum on Forests (IFF) between climate change and biodiversity. *International Environmental Agreements: Politics, Law and Economics* 1: 447–68.

Rowlands, I. H. 2001. Transnational corporations and global environmental politics. In *Non-state Actors in World Politics*, edited by D. Josselin and W. Wallace. New York: Palgrave, pp. 133–49.

Sands, P. 1994. *Principles of International Environmental Law: Frameworks, Standards and Implementation*, vol. 1. Manchester: Manchester University Press.

Scarff, J. 1977. The international management of whales, dolphins and porpoises: An interdisciplinary assessment. *Ecology Law Quarterly* 6: 323–571.

Schweder, T. 2000. Distortions of uncertainty in science: Antarctic fin whales in the 1950s. *Journal of International Wildlife Law and Policy* 3 (1): 73–92.

Schweder, T. 2001. Protecting whales by distorting uncertainty: Non-precautionary mismanagement? *Fisheries Research* 52: 217–25.

Scruton, R. 1996. *A Dictionary of Political Thought.* Basingstoke: Macmillan.

Selin, H., and S. D. VanDeveer. 2003. Mapping institutional linkages in European air pollution politics. *Global Environmental Politics* 3 (3): 14–46.

Sharp, P. 1999. For diplomacy: Representation and the study of international relations. *International Studies Review* 1 (1): 33–57.

Short, N. 1999. The role of NGOs in the Ottawa process to ban landmines. *International Negotiation* 4 (3): 481–500.

Simons, M. 1994. Nations sign pact to stop desert growth: Accord aims to save land for agriculture. *New York Times*, 16 October.

Sjöstedt, G. 1994. Issue clarification and the role of consensual knowledge in the UNCED process. In *Negotiating International Regimes: Lessons Learned from the United Nations Conference on Environment and Development*, edited by B. I. Spector, G. Sjöstedt, and I. W. Zartman. London: Graham and Trotman, pp. 63–86.

Skodvin, T., and S. Andresen. 2003. Nonstate influence in the International Whaling Commission, 1970–1990. *Global Environmental Politics* 3 (4): 61–86.

Smith, J., R. Pagnucco, and C. Chatfield. 1997. Social movements and world politics: A theoretical framework. In *Transnational Social Movements and Global Politics: Solidarity Beyond the State*, edited by J. G. Smith, C. Chatfield, and R. Pagnucco. Syracuse: Syracuse University Press, pp. 59–77.

Snow, D. A., and R. D. Benford. 1992. Master frames and cycles of protest. In *Frontiers in Social Movement Theory*, edited by A. D. Morris and C. M. Mueller. New Haven: Yale University Press, pp. 133–55.

Spector, B. I., and I. W. Zartman. 2003. Regimes and negotiation: An introduction. In *Getting It Done: Post-agreement Negotiation and International Regimes*, edited by B. I. Spector and I. W. Zartman. Washington, DC: United States Institute of Peace Press, pp. 3–10.

Spencer, L., J. Bollwerk, and R. C. Morais. 1991. The not so peaceful world of Greenpeace. *Forbes Magazine*, 11 November, pp. 174–80.

Speth, J. G. 2003. Perspectives on the Johannesburg Summit. *Environment* 45 (1): 24–29.

Stairs, K., and P. Taylor. 1992. Non-governmental organizations and the legal protection of the oceans: A case study. In *The International Politics of the Environment*, edited by A. Hurrell and B. Kingsbury. Oxford: Clarendon Press, pp. 110–41.

Starkey, B., M. A. Boyer, and J. Wilkenfeld. 2005. *Negotiating in a Complex World: An Introduction to International Negotiation*, 2nd ed. Lanham, MD: Rowman and Littlefield.

Stevens, W. K. 1997a. Battle stage is set. *New York Times*, 23 October, p. A20.

Stevens, W. K. 1997b. The climate accord: The outlook. *New York Times*, 12 December, p. A16.

Susskind, L. E. 1994. *Environmental Diplomacy: Negotiating More Effective Global Agreements*. New York: Oxford University Press.

Susskind, L. E., B. Fuller, M. Ferenz, and D. Fairman. 2003. Multistakeholder dialogue at the global scale. *International Negotiation* 8 (2): 235–66.

Széll, P. 1993. Negotiations on the ozone layer. In *International Environmental Negotiation*, edited by G. Sjöstedt. Newbury Park, CA: Sage, pp. 31–47.

Tapper, R. 2002. Environment business and development group. In *The Cartagena Protocol on Biosafety: Reconciling Trade in Biotechnology with Environment and Development?* edited by C. Bail, R. Falkner, and H. Marquard. London: Earthscan, pp. 268–72.

Tarasofsky, R., ed. 1999. Assessing the International Forests Regime: IUCN Environmental Law and Policy Paper 37. IUCN Publication Services, Cambridge, UK.

Tesh, S. N. 2000. *Uncertain Hazards: Environmental Activists and Scientific Proof.* Ithaca: Cornell University Press.

Tetlock, P. E., and A. Belkin. 1996. *Counterfactual Thought Experiments in World Politics: Logical, Methodological and Psychological Perspectives.* Princeton: Princeton University Press.

Thomas, D. S. G., and N. J. Middleton. 1994. *Desertification: Exploding the Myth.* New York: Wiley.

Tolba, M. K. 1998. *Global Environmental Diplomacy: Negotiating Environmental Agreements for the World, 1973–1992.* Cambridge: MIT Press.

Tønnessen, J. 1970. *Den pelagiske fangst 1937–1969.* Sandefjord: Norges Hvalfangstforbund.

Underdal, A. 1989. The politics of science in international resource management: A summary. In *International Resource Management: The Role of Science and Politics*, edited by S. Andresen and W. Østreng. London: Belhaven Press, pp. 253–67.

UNEP and WMO. nd. *Common Questions about Climate Change.* Nairobi, Kenya: UNEP.

UNEP. 1994. *Convention on Biological Diversity: Text and Annexes.* Geneva: UNEP.

UNEP. 2000. *Cartagena Protocol on Biosafety to the Convention on Biological Diversity.* Geneva: UNEP.

UNEP. 2002. *Civil Society Consultations on International Environmental Governance, 22–23 May 2001.* United Nations Environment Programme, 2001. Available at ⟨http://www.unep.org/IEG/Meetings.asp⟩.

UNFCCC Secretariat. 2006 *Kyoto Protocol: Status of Ratification* UNFCCC Secretariat. Available at ⟨http://unfccc.int/files/essential_background/kyoto_protocol/application/pdf/kpstats.pdf⟩.

United Nations Environment Programme. 1998. *Report on International Scientific Advisory Processes on the Environment and Sustainable Development.* Nairobi: UNEP/DEIA/TR.98-1.

United Nations General Assembly. 1992. Resolution 47/188, 22 December.

United Nations. 1980. *Yearbook of the United Nations 1977.* New York: United Nations Department of Public Information.

United Nations. 1983. *International Tropical Timber Agreement, 1983.* New York: United Nations.

United Nations. 1992a. *Agenda 21: The United Nations Program of Action from Rio*. New York: United Nations Department of Public Information.

United Nations. 1992b. *Convention on Biological Diversity*. New York: United Nations.

United Nations. 1992c. *Non-legally Binding Authoritative Statement of Principles for a Global Consensus on the Management, Conservation and Sustainable Development of all Types of Forests*. New York: United Nations.

United Nations. 1992d. Report of the United Nations Conference on Environment and Development (vols. 1–3). Rio de Janeiro: United Nations. UN/A/CONF.151/26/Rev.1.

United Nations. 1992e. *United Nations Framework Convention on Climate Change*, UNFCCC Secretariat. Available from ⟨http://unfccc.int/essential_background/convention/background/items/2853.php⟩.

United Nations. 1994a. *Convention to Combat Desertification in Those Countries Experiencing Serious Drought and/or Desertification, Particularly in Africa: Text with Annexes*. Geneva: UNEP's Information Unit for Conventions.

United Nations. 1994b. *International Tropical Timber Agreement*, 1994. New York: United Nations.

United Nations. 1995. UN document E/CN.17/IPF/1995/2. Commission on Sustainable Development, Programme of Work of the Intergovernmental Panel on Forests, 16 August.

United Nations. 1997a. *Kyoto Protocol to the United Nations Framework Convention on Climate Change* UNFCCC Secretariat. Available at ⟨http://unfccc.int/resource/docs/convkp/kpeng.pdf⟩.

United Nations. 1997b. UN document E/CN.17/1997/12. Report of the Ad hoc Intergovernmental Panel on Forests on its fourth session (New York, 11–21 February 1997), 20 March.

United Nations. 1997c. UN document E/CN.17/IPF/1997/6. Results of the International Meeting of Indigenous and Other Forest-Dependent Peoples on the Management, Conservation and Sustainable Development of All Types of Forests (Leticia, Colombia, 9–13 December 1996), 17 January.

United Nations. 2000. UN document E/CN.17/2000/14. Report of the Intergovernmental Forum on Forests on its fourth session (New York, 31 January–11 February 2000), 20 March.

United Nations. 2001. Information Note 1, Suggestions for a Multi-year Programme of Work of the United Nations Forum on Forest (sic), 1 February.

United States Government. 1991. *America's Climate Change Strategy: An Action Agenda*. Washington, DC: Government Printing Office.

Verolme, H. J. H., and J. Moussa. 1999. *Addressing the Underlying Causes of Deforestation and Forest Degradation: Case Studies, Analysis and Policy Recommendations*. Washington, DC: Biodiversity Action Network.

Vogler, J. 2003. Taking institutions seriously: How regime analysis can be relevant to multilevel environmental governance. *Global Environmental Politics* 3 (2): 25–39.

Willetts, P. 1996a. Consultative status for NGOs at the United Nations. In *The Conscience of the World: The Influence of Non-governmental Organisations in the UN System*, edited by P. Willetts. London: Hurst, pp. 31–62.

Willetts, P. 1996b. From Stockholm to Rio and beyond: The impact of the environmental movement on the United Nations consultative arrangements for NGOs. *Review of International Studies* 22: 57–80.

Williams, M., and L. Ford. 1999. The World Trade Organisation: Social movements and global environmental management. *Environmental Politics* 8: 268–89.

World Rainforest Movement. 2000. *Report on the 8-Country Initiative: An International Expert Meeting to Discuss "Shaping the Programme of Work for the United Nations Forum on Forests (UNFF),"* Bonn, 27 November–1 December 2000. Available at ⟨http://www.wrm.org.uy/actors/IFF/8country.html⟩ (accessed 7 January 2003).

World Rainforest Movement. 2002. *UNFF: Little Hope for Forest Peoples and Forest Biodiversity in this Forum.* Available at ⟨http://www.wrm.org.uy/bulletin/56UNFF.html⟩ (accessed 7 January 2003).

World Resources Institute. 1997. *US Competitiveness Is Not at Risk in the Climate Negotiations.* Washington, DC: World Resources Institute.

Yamin, F. 2001. NGOs and international environmental law: A critical evaluation of their roles and responsibilities. *Reciel* 10 (2): 149–62.

Yearley, S. 1994. Social movements and environmental change. In *Social Theory and the Global Environment*, edited by M. Redclift and T. Benton. London: Routledge, pp. 150–68.

Young, O. R. 1997. Rights, rules and resources in world affairs. In *Global Governance: Drawing Insights from the Environmental Experience*, edited by O. R. Young. Cambridge: MIT Press, pp. 1–23.

Young, O. R. 2002. *The Institutional Dimensions of Environmental Change: Fit, Interplay and Scale.* Cambridge: MIT Press.

Zartman, I. W. 1993. Lessons for analysis and practice. In *International Environmental Negotiation*, edited by G. Sjöstedt. Newbury Park, CA: Sage, pp. 262–74.

Zartman, I. W., and M. R. Berman. 1982. *The Practical Negotiator.* New Haven: Yale University Press.

Zürn, M. 1998. The rise of international environmental politics: A review of current research. *World Politics* 50: 617–49.

Zürn, M. 2004. Global governance and legitimacy problems. *Government and Opposition* 39 (2): 260–87.

Index